A SCIENTIFIC LIFE

GRAHAM RICHARDS

authorHOUSE®

AuthorHouse™ UK
1663 Liberty Drive
Bloomington, IN 47403 USA
www.authorhouse.co.uk
Phone: UK TFN: 0800 0148641 (Toll Free inside the UK)
UK Local: 02036 956322 (+44 20 3695 6322 from outside the UK)

Published by AuthorHouse 01/14/2021

ISBN: 978-1-6655-8443-2 (sc)
ISBN: 978-1-6655-8444-9 (hc)
ISBN: 978-1-6655-8441-8 (e)

Print information available on the last page.

This book is printed on acid-free paper.

Cover image created by Dr. Jane Burridge for the cover of the Journal of Molecular Graphics

CONTENTS

PREFACE

Why would anyone be interested in my life? The proverbial 'good question' as one often responds when initially stumped for a clear answer. The fact is that I have had a reasonably successful scientific life. I ended up as the first head of the chemistry department at Oxford, by some measures the biggest chemistry department in the world and one of the most distinguished. This has been recognised with a series of awards, some trivial but at least a couple of significance. However, what makes my career rather different from that of many academic scientists is that I was particularly involved in the creation of high-tech companies based on university intellectual property. When I started in that direction, this was very non-standard and even frowned upon. Now, of course, it is both commonplace and fashionable. It is in this respect where my experience may be of assistance to younger colleagues, and I hope of more general interest. At the very least, it should be of some help to the poor devils who will have to write my obituary for the Royal Society.

1

ORIGINS

Being born within days of the outbreak of World War II inevitably meant a somewhat disrupted childhood. Although born at home in Greasby on the Wirral Peninsula, my origins were very Welsh. My mother, one of fourteen children in deeply rural mid-Wales, left school at the age of eleven, even though the legal age to do so was fourteen. Where they lived was too far from any secondary school. Only her two youngest brothers managed to have secondary educations thanks to the generosity of the Davies family, of whom more is told later. As a teenager, she was sent off to England to work as a servant to a wealthy family in Birkenhead, where one of her sisters was the cook. My father's background was similar. His mother had gone from childhood in Rhosllanerchrugog to marry my grandfather, who worked for the Great Western Railway in Ruabon but was then transferred to Birkenhead, in those days the northern end of the GWR. The stationmaster was the father of the poet Wilfred Owen.

Just at the time of my birth, my father, a printer, became the managing director of a printing company in Leeds, so we moved to the village of Horsforth, near Leeds. The outbreak of war meant that the printing trade vanished as there was no paper. My mother had severe postnatal depression and attempted suicide.

The family solution was to go back to mid-Wales, where there was a clan of supportive relatives. We lived first with my aunt F at Gregynog Hall, where she was the cook. I then, as a child, met the benevolent Davies sisters, who had a wonderful collection of impressionist paintings, including Renoir's *La Parisienne*, that I remember as the girl in the blue dress. It, along with their other paintings, are now in the Welsh National Museum.

I first went to school in the neighbouring village of Tregynon from Gregynog, but soon we moved in with another aunt, Sal, who lived with the Rector and Mrs Richards in Newtown rectory. They had essentially adopted Sal, largely as a servant, but she stayed with them for over seventy years. Both these aunts were unmarried and wonderfully kind to me, very much second mothers. While in Newtown I got to meet and know most of my twenty-nine first cousins. Somewhat later, the fact that three of them became university lecturers certainly influenced me, in particular my cousin Mervyn, who was a physicist, and Eric, who had a distinguished war career as a flight engineer in

Bomber Command. He, like many of my relatives, was a natural engineer. He forged his age to join the Royal Air Force (RAF) and flew with the 617 Squadron of Dam Busters fame. Initially he was denied a commission as he had no school certificate, but after many bombing raids, he was promoted, finishing the war as a squadron leader engineer. With the advent of peace, most had to leave the forces, but with the introduction of jet engines, Eric was asked to stay on. However, once the war was over, the fact that he had no school certificate became once more problematic, so he left and became a bus driver, although later running a garage.

With peace, just after VJ Day we were rather pushed out of the rectory and wanted to return to Greasby, where my parents still owned a house, but it was let to a tenant family. Just as my parents were trying to return to their house, the husband of the tenant family died, causing them financial loss. As a result, our family moved in alongside our tenants. It was rather cramped in a standard British two-up, two-down, semi-detached house.

This crowding became even more of a problem when, in the summer of 1946, I went down with polio, then generally called infantile paralysis. I was hospitalised in Birkenhead and have vivid memories of my time there in the dreadful snowy winter just prior to the creation of the National Health Service (NHS). In that era, the seemingly bizarre belief of the medical profession was that it was damaging for children in hospital to be visited by their parents. Thus for eight weeks I did not see my parents and was admonished by the nursing sister for writing upsetting letters to my mother, wondering if I would ever see her again. This was made all the worse by being kept in isolation for the first few weeks, but it did wonders for my reading. To make matters even more fraught, I needed an iron lung, but the hospital did not possess such a thing. Fortunately for me, the Royal Navy (RN)—in which polio had been rife, particularly in the submarine service—provided one for me. They also sent me some illustrated books about the war which I read many times. I still remember them in some detail. Happily, I recovered largely from polio apart from one slightly withered leg, and the experience probably made me more keen on sport and fitness than would otherwise have been the case.

My primary school was not very good and was totally focussed on the eleven-plus exam. At that time the Wirral had, and indeed still does have, excellent state grammar schools. I passed the eleven-plus with a place at the excellent Calday Grammar School. However, in that era there was an even more prestigious option. This was still the time of direct-grant schools, a brilliant option later abolished by the wretched Shirley Williams. Birkenhead School was at that time academically superb, and I tried their entrance exam. I succeeded in the top batch, therefore not requiring an interview. I was only the second boy from my primary school to gain entry.

Birkenhead School was superb, and I thrived there, almost always top of the class and doing very well in all subjects except English. My spelling was, and is, awful. A key decision had to be made when one went into the senior school at the age of thirteen. At this stage, the top form had to choose their third language, having done Latin and French in the junior school. The choice was between Greek and German. When the option was presented, almost all opted for German. But then the very able classic masters got to work, and at least half, including the brightest boys, were persuaded to study Greek. Even at that stage I knew I wanted to become a scientist and opted for German.

In that era, for a short mad time, one was not allowed to take O-level exams before the age of fifteen. Because of that and the headmaster's keenness to get pupils into Oxford and Cambridge, which generally required three years in the sixth form, the top three boys in the fourth form were jumped past the fifth form, where O levels were taken, straight into the lower sixth. To satisfy Oxbridge entry requirements, I just took three O levels while starting in the lower sixth: Latin, German, and English. In jumping the fifth form, I missed out on the teaching of calculus, but fortunately for me, my very able cousin Alan taught me this topic over the holidays, and I was very well prepared. He became a university lecturer in Swansea and a pioneer in computing.

As to this day, the crucial choice of A-level subjects has to be taken at a very early stage, when pupils are uncertain about their futures and much influenced by inspiring teachers. I opted for physics and double maths, being pretty sure that my future career would be in physics. Unfortunately, in that year, only three of us made that choice, so the option was removed, and

I had to do A-level physics, maths, and chemistry. The physics master, Frank Ellis, was very inspiring, but the chemistry teachers anything but.

I managed to gain a state scholarship as a result of my A- and S-level marks in my second year in the sixth form which was unusual. Then I went back to school for a third year in the sixth form, standard practice at the time. The upper-sixth form was essentially for potential Oxbridge candidates. I applied for natural sciences at Cambridge but was rejected, although it was suggested that I could have a place to read mathematics. I knew that I was not a mathematician and so applied, as one then could, to Brasenose College in Oxford to read physics. I received a letter offering me a place but to read chemistry. To this day I don't know whether this was prescience on behalf of the Brasenose dons or an administrative error. I was too naïve to question this decision and so accepted the place to read chemistry, which I had not really cared for.

2

STUDENT DAYS

All generations of students think they are special and possibly unique. Those of us who went up to Brasenose College in Oxford in 1958 can justify that claim better than most, particularly if that class includes, as is reasonable, those who came up in 1959 but went into the second year and hence took their finals with most of us—the class of 1961, in the North American usage, which dates by the year of graduation rather than of matriculation. The most notable additions were the several Rhodes Scholars.

One unique feature was that 1958 was the year when national service was quite suddenly stopped. In my own case, I had expected to do the two years of compulsory military service between school in Birkenhead and coming up to Oxford. In that era, the School Cadet Force, the CCF, was quite a serious business. One tried to do well in the Corps so as to optimise the chance of being commissioned as an officer during national service. In my own case, I even went on a course with the Royal Marine Volunteer Reserve on the Mersey in the hope of doing my stint in the marines. In general, most national servicemen did their time in the army or the RAF. Getting into the navy or marines was less straightforward. However, if one did go into the marines, the chances of being commissioned were much improved.

A number of my schoolmates who did get into the Royal Navy (RN) were then presented with the dilemma of being offered a commission if they were prepared to fly. For some of my friends, this was blocked by strong parental opposition. They spent their two years as able seamen, often going as the funeral parties of those who took the flying option, where casualties were frequent, particularly in the early days of the Buccaneer as a carrier-based plane.

Being a tiny minority of lowly paid national servicemen amongst regular matelots with much better pay also resulted in some of them becoming very good bridge and particularly poker players. After all, there is not much to do for most of the day on a warship. When they finally came to Oxford, some of my school friends with this background became almost professional gamblers. They sometimes did not need to cash their grant cheques for weeks. Others went hollow-eyed round the streets, begging off friends because they had to have several hundred pounds by the following Friday.

Suddenly in 1958, the Conservatives most improbably

abolished conscription. Young men had been called up by quarters. Those born in the first two quarters of 1939 were liable; the third quarter was less clear-cut. But those born in the fourth quarter of 1939 were thereafter not legally liable. I was born on the first day of that fourth quarter, so to my surprise and that of my college, I was free to come up to Oxford without doing my spell in the forces.

The sudden cessation posed a problem for the colleges. They had most of a year of freshmen accepted who would be coming out of the army and another, including me, who were straight from school. Some had to find something to do for a year, a precursor of the subsequently popular and valuable gap year. I was allowed to come up immediately but with the proviso that there was no college accommodation for me. In fact, I spent my first year living in The Station Hotel, now The Westgate. It was a small commercial hotel run by Miss Johnstone, who had been a traditional university landlady and liked to let three of the rather small rooms to undergraduates. Through this I met some interesting transient guests, including the cricket umpires who used to stand in The Parks. Most usefully I managed to keep a scooter which greatly benefitted my social life.

The Oxford of 1958 was a great mixture of those who had and had not done military service. Social class was very much emphasised by those who had been commissioned and those who had not. And within that classification, some regiments or experiences counted more than others. Ex-naval officers or RAF pilots counted highly, as did those who had been Guards,

Officers, or the like. Many scientists had been in technical corps, a slightly lesser breed.

One particularly interesting set were those who had the good fortune to spend their two years of service learning Russian. All the RAF and RN Russianists were commissioned, while those in the army might be or perhaps remain as sergeants, even doing the same course. It must have been a wonderful opportunity and had a significant influence of British letters and academia. Those ranks included Alan Bennett, Michael Frayn, Dennis Potter, and several of my subsequent colleagues among the Brasenose Governing Body.

Largely because of the leavening of ex-national servicemen, the undergraduates of the late 1950s had a direct link with the post-war generation, the bulk of whom had served in the war. Many of my contemporaries had seen action in Korea, Suez, Cyprus, Aden, Kenya, or Malaya. Certainly as a body, they were significantly more mature and self-reliant than current undergraduates. We wore sports coats and ties and often cavalry-twill trousers. As a group we looked pretty indistinguishable from any year since 1919. The big change came after us, with national service a thing of the past. Undergraduates became students with long hair and irreverent attitudes, challenging authority and listening to the Beatles.

Even though the average age was much higher than today, and the fact that many of my contemporaries had seen active service, the college gate was locked at ten at night, after which one was expected to climb in. Keys were not provided for undergraduates until 1967, in fact first by me, when I was dean of Brasenose.

The Oxford Colleges of the 1950s were much more differentiated than is now the case. Balliol was the intellectual and political college; Christ Church had a definite upper-class persona. Brasenose was still, as it had been for many years, the pre-eminent sporting college in the university. The class of '58 was again unique in that it represented the Indian summer of that tradition. No group in subsequent years in that or any other college has come close to matching our athletic achievements. The class produced a rowing blue; a rugby blue; and two athletics blues, including the president of the University Athletics Club. We earned two football blues, including the captain; two hockey blues; four lacrosse blues, including the captain; and a cricket blue, who went on to play county cricket. Members of this vintage crop took part in teams which won the inter-college cup at rugby, athletics, and hockey.

It might be thought that such a concentration on sport would bode ill for the academic achievement of the group. But in fact, the record in final exams was more than adequate, and the subsequent careers have been marked by singular success. Out of about a hundred freshmen have come a high-court judge, a senior partner of a big-four city law firm, two Hollywood moguls, a university chancellor, and two generals. There was a trades union general secretary, a senior press and a senior TV journalist, numerous top-level businessmen, and several university professors. The freshmen class also produced a headmaster, a Fellow of All Souls, and a Fellow of Brasenose. Just one ended up in jail.

It is certain that some of this group had been originally selected on grounds of their sporting prowess rather than single scholastic scores. That would never happen these days. Student admissions are bedevilled with excessive bureaucratic regulations which do not permit tutors to back their judgement on selecting pupils. However, some of the qualities of a non-academic nature may be far better predictors of future eminence than A-level grades.

Life as an undergraduate in the late 1950s was much less organised than is now the case. We normally had one tutorial per week at which an essay had to be presented or read, but not the plethora of classes and revision sessions which are now common. Nor were we constantly examined; even informal college exams, or 'collections' were not heavily used. I read chemistry, that was anomalous as a subject in a variety of ways. In those days, there were just two exams during one's course—the first public examination, or prelims, and then the second public examination, finals, given at the end of the third year. Chemists usually took their prelims before coming into residence. This meant coming up to Oxford prior to starting the course to take the exam.

As in all Oxford public examinations, one is required to wear so-called 'sub-fusc'—a dark suit, white bow tie, gown, and cap. However, military uniform, according to the rules, counted as sub-fusc. I well remember the man in front of me in my prelim exams was wearing the dress uniform of the military police.

With that out of the way before starting the first year, I had no more examinations until the end of my third year, when there was essentially a one-word syllabus—'chemistry.' We could be asked whatever the examiners chose to ask. And neither the lecture course nor the tutorials were coordinated with each other, nor with the final exams. In many ways this was pretty chaotic, but it did give undergraduates time to think and to chase up things which particularly interested them.

It was also possible to get away with relatively little work, especially if one was well-organised. I was not a particularly diligent student. I spent most afternoons involved in some sport or another and much of my evenings at parties. There were very few women undergraduates, so female company was provided by nurses, girls at the secretarial schools, and by some of the language schools catering to continental European students. The latter was my particular hunting ground, and I suspect that converted me into a lifelong Francophile.

The second unique feature of chemistry at Oxford was, and is, that it is a four-year course with final examinations at the end of year 3. This was followed by a year of research, which constitutes part II of the course. After finals, one had a fair idea of how well one had done, so I started my part II knowing I was likely to get a first-class degree, which in those days covered about 5 per cent of the candidates; now it is about 30 per cent. Grade inflation is another rule in education. I opted for my part II year to work in the physical chemistry laboratory under the supervision of my tutor, Richard Barrow.

3

PHYSICAL CHEMISTRY LABORATORY

My research life started when I began my undergraduate fourth-year part II project in Oxford's physical chemistry laboratory. The PCL, as it was then known, had been completed as a new university laboratory in 1941 funded by Lord Nuffield. It replaced the research facilities housed in the college laboratories of Balliol and Trinity Colleges. The creation of the new laboratory was achieved by its head of department, Cyril Hinshelwood, who in 1936 had succeeded Frederick Soddy. Soddy had a very distinguished career. He, along with Rutherford, explained radioactivity. Later, he proved the existence of isotopes, for which he received the 1921 Nobel Prize for Chemistry. In his later years, however, in the period 1921 to 1934, he turned his interests to economics and was largely treated as a crank; now, with the benefit of hindsight, he is considered to have been a visionary. When I joined Hinshelwood's department, the PCL had eight what we would now call professors heading the independent research groups. Of the eight, six were Fellows of the Royal Society and included in that era its president, a council member and its foreign secretary.

Hinshelwood, himself a Nobel Prize–winner, was also in control of the nearby inorganic chemistry laboratory, later a separate department. Its active scientists included Dorothy Hodgkin, later Britain's only female scientific Nobel laureate. These days the name of Hinshelwood is essentially unknown, even amongst scientists. Yet he was probably one of the great intellects of the twentieth century. Not just a Nobel Prize–winner, he also had a gift for languages. He spoke and wrote French, German, Italian, and Spanish perfectly, and knew Russian and Chinese plus had an acquaintance with Arabic. As well as being president of the Royal Society in its tercentenary year, he served as president of the Classical Association and the Oxford Board of the Modern Language Association. He was a significant scholar of Dante. Despite this learning, he was a somewhat solitary figure who viewed the outer world with distrust, not exactly loved by his colleagues, but nonetheless generous with students.

Generally known as 'Hinsh,' he was particularly intolerant of university bureaucracy, where he felt he had to defend the sciences against the central powers. This caused his colleague Hammick to pen the verse:

Straight to the throne of God he strode,
Cyril, first Baron South Parks Road.
Swore that all should go to hell,
Unless they worked in the PCL.
But the Good Lord would not condemn
For 'They' had been got at by 'Them.'

Educated at Westminster City School, Hinshelwood won a scholarship to Balliol College, Oxford. He was unable to take this up immediately due to the First World War, and from 1916 to 1918, he worked at the Department of Explosives Queensferry Royal Ordinance Factory. His ability was immediately recognised as at this tender age he became deputy chief chemist of the main laboratory. His work there on the slow decomposition of explosives, by measuring the gas evolved, stimulated his interest in chemical change, the topic of much of his subsequent work.

In 1919 he went to Balliol to do the shortened post-war course in chemistry with his tutor, Harold Hartley—later Sir Harold—who, as a brigadier-general, had been the controller of chemical warfare. Regrettably, Hartley had forgotten much of his chemistry during the war and subsequently moved to have a very distinguished career in business and industry, including serving as chairman of the British Overseas Airways Corporation. Hinshelwood worked on gas kinetics with his work on chain reactions leading to the Nobel Prize, that he shared with the Russian Semenov.

Also concentrating on chemical kinetics, but not particularly collaborating with Hinshelwood was R. P. (Ronnie) Bell, who had succeeded Hinsh as the Balliol physical chemistry tutor. He again did very distinguished research and was unlucky not to win a Nobel Prize for his explanation of the role of proton tunnelling in reactions.

When I started in the laboratory, the presence of Sir Harold Thompson loomed large. This was not just because of his distinction as the leading figure in infrared spectroscopy, but rather his eminence in this world of football. He had founded the hugely successful Oxford and Cambridge side Pegasus which won the Amateur Cup in the era when that was a huge event. He went on to become chairman of the Football Association. In addition to these giants, also working in the PCL were E. J. Bowen, the father of photochemistry, and L. E. Sutton, who made slightly less impact but was nonetheless one of the most cited authors in chemical publications.

Slightly younger, but no less significant, was Rex Richards (no relation). He was a major factor in the topic of nuclear magnetic resonance (NMR) being taken up by chemists. Then later in his career he pioneered the application of NMR to problems in molecular biology. He succeeded Hinshelwood as the head of department's chair, the Dr Lee's Professorship, and after that became vice chancellor of Oxford University.

I joined the department in 1961 to work under the supervision of R. F. Barrow, who had been my undergraduate physical chemistry tutor. Richard was a pretty awful tutor. The weekly hour was mostly silence while he looked for an ashtray before lighting his second cigarette. By contrast, he was a wonderful research supervisor, fully engaged in his students' research down to the finest details. His area of research was the study of the electronic spectra of diatomic molecules, of which he was one of the world's foremost proponents.

My project was to work at the electronic absorption spectrum of chlorine, particularly the higher vibrational levels so that we could get a clearer understanding as to why the dissociation energy of the molecule is so much higher than that of fluorine and bromine. Having obtained values of the vibrational levels close to the point where the two atoms fall apart, it made sense to draw the potential energy curve showing how the energy of the molecule varies as a function of the distance between the two atomic nuclei. There was a standard method to do this, the so-called Rydberg-Klein-Rees method. To use this formulation, one had to do some difficult and tedious integrals.

This is one of several points when luck played a crucial role in my life. I realised, perhaps out of laziness, that there was this newfangled thing called a computer which could do integrals numerically. I was thus one of the earliest chemists to use a computer. In my research group no one had taken this step; the conventional view was that they were unlikely to be much use to chemists. It must be remembered that at that time they were, by modern standards, rather primitive. The Oxford University computer was a Ferranti Mercury machine. It was the size of a couple of rooms, had a 32K memory, filled with valves, programmed in autocode or machine language, and communicated with by paper tape. The queue was literally a queue, where one stood in line clutching one's paper tape. When you reached the tape-reader, it would read until a mistake was found, such as, "Error 31," when trying to take the square root of a negative number. One then marked the spot where the error showed up, rushed to a teleprinter, typed a corrected version, cut and spliced the tape, and rejoined the queue. Errors were only found one at a time.

Despite that primitive nature of the machine, it served me well. But more significantly, it enabled Dorothy Hodgkin to solve the crystal structure of penicillin and other important molecules. In this she was assisted by very bright young men and women, most of whom were not at the time regarded as at the forefront of research.

In my case, by the time I submitted my doctoral thesis, computing was starting to take off. Sadly it was a classic British story. We did the initial, often groundbreaking, work, but turning this into commercial success eluded us. The ideas were taken up by foreign companies and built into international companies. One can cite similar stories for dye stuffs, television, jet aircraft, and antibiotics.

The Ferranti Mercury was replaced in Oxford by the English Electric KDF9 machine. Out at Harwell, the same company produced a big machine, Atlas. Despite this start, in no time IBM had taken the market, and the British computing industry disappeared.

For me, writing a chemistry doctoral thesis in 1964 which was largely about computing did, however, at least make me look clever. And it made it clear to me that I wanted to do research. My main examiner was Rex Richards, who later was very good to me. The next obvious step was to apply for a college junior fellowship, a time-limited opportunity to concentrate on research.

4

JUNIOR FELLOW

While finishing my doctorate in 1964, I applied for junior research fellowships at three Oxford colleges. Postdoctoral opportunities were much less available in those days. I was lucky and invited for interview at Balliol, Merton, and my undergraduate college, Brasenose. The interviews were all in the same week, so I withdrew my Merton application and was interviewed by Brasenose and Balliol. The Balliol interview was a little bizarre. The master, Sir David Lindsay Kier, opened the questioning by asking where I lived. I replied that I lived in a flat in Oxford which I shared with a Balliol man, George Alberti (later president of the Royal College of Physicians). "Ah, dear George," said the master. "Were you at George's wedding?" We then discussed the wedding for most of the time scheduled for the interview and the obvious scientific questions. Ronnie Bell, Heine Kuhn, or Pat Sanders barely got a question in. Despite this I was offered the fellowship, that was at no cost to the college since I had already won an ICI Research Fellowship which paid my salary.

Hence, in October 1964, I moved into Balliol and started independent research. The college at that time was the preeminent college in academic terms and a very exciting place to live. As I have mentioned, the scientists there included Bell, Kuhn, and Sanders, as well as the young and exciting Malcolm Green. The history dons were particularly stimulating, and I much enjoyed the company of Christopher Hill, Jack Gallagher, Richard Cobb, and Maurice Keen.

In the laboratory I continued in the area of diatomic spectroscopy but branched out by being one of the first people in the UK to obtain an argon ion laser; lasers were very new and definitely the flavour of the moment. At the same time, I moved more in the direction of becoming a theoretical chemist using computational methods. In this latter domain, one problem was the view held by Charles Coulson, the preeminent Oxford theoretician, and his former pupil, Christopher Longuet-Higgins, his Cambridge opposite number. Both were very distinguished, respected, and doing brilliant work. But both were also temperamentally hostile to the use of computational methods as opposed to elegant paper theory. Their influence held back the area for some years.

I gave a seminar about my work on potential energy curves and their computational calculation which was attended by the Paris-based American Carl Moser, a former Coulson pupil.

Carl liked what I was doing and invited me to come and spend a year in his lab in Paris. Despite being warned against this by Coulson, and the fact that I had been discussing a period at Harvard, I accepted Carl's offer, above all as by this time the computational facilities in France were vastly superior to those to which I had access in England.

There was one problem. My ICI Research Fellowship, that paid my salary, was not transferable. Fortunately for me, Hinshelwood, in his last days as professor, came to my aid. Although Hinsh hated government committees, he enjoyed working with industry. Most notably he appreciated his association with ICI, both the Alkali Division at Northwich and Dyestuffs at Blakeley. Hinsh managed to get the ICI rules changed so that I could draw my ICI Fellowship salary while in Paris at the Centre de Mecanique Ondulatoire Appliquee (Centre of Applied Wave Mechanics). Even more fortuitously, on getting to Paris, I found that Carl had also arranged for a salary for me from the CNRS, the French Research Organisation (Centre National de Recherches Scientifique). This meant I had two salaries and left very rich, although the French salary was somewhat peculiar. Each month I had to visit personally the headquarters of the CNRS on the Quai in St Germain to ask for the money. Every time they denied all knowledge of me until I finally got through to a Mademoiselle Valentin, who would pay me in notes. For some months I felt unable to spend the cash. But when I did, I lived very well, moving from a room in the Cite Universitaire to a *chambre de bonne* in rue des Saints Peres on the Left Bank.

Immediately before going to France, I had spent a couple of months in Stockholm, where my former supervisor, Richard Barrow, was having a sabbatical. One aspect of working in Sweden was that everyone spoke English. And what is more, when I went into a lab, out of politeness the other researchers would switch from Swedish to English. The only Swedish I learned was how to swear. By contrast, in the French lab, even though many of my colleagues had been postdocs in the United States and could speak excellent English, they chose not to do so and only spoke French. I initially found this rather unfriendly, but after a few months, I appreciated this approach since it meant that I learned to speak French.

The laboratory was in the rather left-wing 19th Arrondissement with the nearest Metro station, Stalingrad, Avenue Jean-Jaures close by, as well as the then flourishing abattoir La Villette (now a museum). The latter meant wonderful food in the local restaurants.

I shared an office with another postdoc, Georges Verhaegen, later the rector of the Free University in Brussels. Together we worked on *ab initio*_molecular orbital calculations using a program written in the IBM lab in San Jose by Bob Nesbet. The program was real tour de force, but entering the starting data was very complex. Some bits had to be entered, for instance, in hexadecimals. Carl Moser kept power by being the only one who could enter the data to run a calculation. Georges and I, however, after some trial and error, cracked the code. My spectroscopic background enabled me to think up a suitable problem to solve. It concerned the potential energy curves for electronic states of the diatomic molecule beryllium oxide, BeO.

We had masses of computing time. This was due to Carl's skill and idiosyncratic behaviour. One of his many quirks was having a Corgi dog, Gunter, which was paraplegic. His back legs did not work, so when taking him for a walk, the dog's collar was put around its waist and the rear end held up with the lead. In addition, Gunter had no control of his bladder, so when taking him for a walk, you had to be careful not to be sprayed. Carl would take Gunter into the government minister's office when demanding a thousand hours of computer time. Fearful of damaging his carpet, the minister would agree. Carl himself had no idea how much a thousand hours were, so initially we had to run the same calculation time after time to use up the allotted ration. But as our use expanded, we were very well served using the IBM 704 machines based out at Saclay. So good were these facilities and superior to what was available in Oxford, that when I returned to Oxford, I obtained a grant from the Royal Society to enable me to go to Paris once a month to run big calculations. By this time, one communicated with computers via punched cards. I used to fly to Paris on a Friday evening and enter my job, usually a very thick wodge of cards. Next morning I would correct any errors and resubmit, returning to London on the Sunday evening after a delightful weekend in France.

In the July of 1966, I returned to Balliol. This was a mad period in British academic history. In the early and mid-1960s, the existing universities doubled in size, the government created several new universities all at the same time. This meant that there were suddenly a vast number of new posts available. Jobs grew on trees, including in Oxford. I was actually offered a college fellowship while still a Balliol junior fellow at another Oxford college. So easy it seemed to get jobs that I rejected this offer as I did not want to be tied down with all the inevitable teaching at such a young age and with my research thriving. None of us appreciated the obvious fact that, once the glut of jobs was filled, there would inevitably be a long period with no new jobs, as did happen. Luckily for me, although I did not appreciate it at the time, my own undergraduate college, Brasenose, then also offered me a fellowship without any advertisement or even an interview. With somewhat ill grace I accepted the offer on condition that I could spend the year in Paris, although missing out on the proposed Harvard stint with Bill Klemperer.

I started as a fellow at Brasenose in 1966. Luckily for me, Rex Richards, by this time the head of physical chemistry, gave me a university lectureship. I had been so ignorant that I had not realised that a college fellowship without a university lectureship would not have been permanent given that then and to this day, Oxford jobs are joint appointments between the largely independent colleges and the university, which in the case of the sciences, the university providing the major part of the salary.

5

LECTURER

Starting out on my own with an independent research group gave me the freedom to pick my own area of research. I had in principle a blank canvas, providing I did something different from the research being done in the same department by my previous research supervisor. Rex Richards encouraged me to have a look into shock waves. I did follow this up by reading about the topic but instead decided to go into the area of lasers. How different that era was than the present, when newly appointed academic staff have often done several postdocs and are committed to a given field of research. Rex's instruction to me was not to be in a hurry and to give some thought about what I would like to do.

Initially I shared a laboratory in the newly enlarged physical chemistry laboratory with Ted Bowen. He was just about to retire but was a delightful companion and mentor with a fount of comic stories and even poems. He also loved building apparatuses, frequently incorporating old tins and bits of rubbish.

As well as my argon ion laser and an optical spectrograph, I had an image intensifier, then quite novel, obtained from a generous colleague in the department of engineering science. With this we managed to detect what we believed to be the first recording of laser-induced resonance fluorescence in an electronic spectrum with just two spectral lines per vibrational level rather than many thousands. This was written up with the intention of submitting to the prestigious journal *Nature.* In those days, long before personal computers, one wrote a manuscript longhand and then had it typed by one of the departmental secretaries. While my paper was waiting to be typed, a group from MIT published essentially the same results. We had been scooped. This happened more than once. Lasers were very much the flavour of the moment, and it was hard for a very small Oxford group to compete with the more lavishly funded and well-established US groups.

By contrast, what started as a sideline, doing ab initio molecular orbital calculations on diatomic molecules seemed to thrive to the extent that I soon abandoned experiments to concentrate on the computational work. At that time, ab initio molecular orbital calculations were very much the province of physicists and applied mathematicians rather than chemists. Keen to demonstrate their power and to encourage chemists, who had the problems to solve, my first graduate student, John

Horsely, and I wrote my first book on the topic. Following on from the work I had done in Paris on the question as to how the molecule BeO would dissociate, given that what was the ground electronic state of the molecule would not break up to give ground-state atoms, we became interested in calculating spin-orbit coupling constants, knowledge of which would clarify the situation. With a brilliant graduate student, Timothy Walker, we developed the software to compute spin-orbit coupling constants. The results were amazingly accurate at a time when many things calculated by ab initio methods were pretty crude.

One aspect of spin-orbit coupling in diatomic molecules is the so-called lambda-doubling, a separation in energy levels so tiny that any transition between them would fall in the radio frequency range of the electromagnetic spectrum. At that time the only molecular species known in interstellar space was the radical OH observed by radio telescopes. There was great interest whether other molecules, particularly involving carbon, might exist in space and thus act as possible building blocks of organic molecules and ultimately life forms. Together with another outstanding student, David Cooper, we computed the lambda-splitting in a carbon-hydrogen compound (CH) and hence a prediction of the frequency at which a radio telescope would detect that carbon-containing compound. The spectrum of CH on earth is very simply observed. Indeed, it is a nuisance as an impurity arising from any bit of carbon around when exciting emission spectra. The great and Nobel-prizewinning spectroscopist Gerhard Herzberg had analysed the spectrum, but inevitably, the excited radical was at a high temperature while interstellar space is cold. Thus to obtain an estimate of the frequency of a transition between the lambda doublets involved a big extrapolation. Even with Herzberg's wide error limits, the search for CH had proved fruitless. However, our theoretical prediction lay outside his limits. This time, while we were waiting to publish our suggested frequency, the radical was discovered almost precisely at the frequency we predicted. Once again our timing was unfortunate, and our paper did not receive much praise. Nonetheless, I still think that work was one of the smartest things I ever did.

Soon after CH was discovered, many organic molecules were detected, including the long, linear polyacetylene cyanide predicted by my friend Harry Kroto. It was important to my group to know for which molecular spectra there had been published ab initio calculations. This being long before modern data scanning software, the only solution was to scan the published literature, going into the library and perusing the contents lists. Since such information was going to be valuable to researchers across the world, together with teams of my research students, we published our findings in book form. These ran to three increasingly fat volumes and covered the period up to 1980. By then, computer bibliographics rendered our approach redundant.

I was lucky in having some very good students. One way of attracting such folk was, and is, to make an impression on undergraduates before they have to make a choice as to with whom to do research. This is achieved by one's teaching. In Oxford there are two parallel sources of teaching. Lectures are given at a departmental level to students from across the whole university, just as in universities across the world. At the same time, at a college level, a tutor gives weekly tutorials to undergraduates, normally two at a time, but occasionally to a lone student.

My first lecture series was specified for me by Rex Richards. He asked me to give a course on statistical thermodynamics. I protested that it was a topic I had struggled with as a student and did not feel that I had a deep understanding of it. 'Just the man then,' said Rex, and he was right. When I made the effort, I found that I did like the topic and felt that I could make it comprehensible. My lecture also became a book which did moderately well. Like my earlier books, it was published by Oxford University Press (OUP), with whom I was to have a complicated relationship.

At the college-level tutorials, I had to cover the whole of physical chemistry, the bits I loved and the bits I disliked. I often think tutors are better at the parts of their subject far removed from their research interests. It can be a hindrance if one knows too much about a topic.

As well as teaching for Brasenose, where we took in from six to eight undergraduates reading chemistry, I also, for many years, taught pupils from Lady Margaret Hall which was, until 1978, all female. This, together with my experience living in the mixed college Francobritannique in the Cite Universitaire in Paris, convinced me of the desirability of having mixed colleges. In my first year as a Fellow of Brasenose, I pushed for this. Being in the mid-1960s meant that the notion of the colleges going mixed was very much in the air. However, to admit women to one of the all-male colleges required a change of statutes which demanded a two-thirds majority of the fellows. A couple of colleges had tried but failed to meet the requirement. Brasenose seemed an unlikely pioneer, but in 1967 I did achieve, by a wide margin, the necessary number of votes. In fact, only three fellows voted against. Armed with the majority, a small subcommittee of Brasenose fellows met with the five heads of the women's colleges, whose collaboration over the complex admission procedures was essential. They, led by Lucy Sutherland of Lady Margaret Hall, were totally opposed. With some justification, their view was that the essential feature of women's colleges was the provision of jobs for academic women. They would not cooperate, but we did manage to extract a promise that, were the subject to arise again, and clearly it would, if any of the all-male colleges were to be permitted to accept women, then Brasenose would be able to. This proved very useful when the issue arose again in 1973, and we were able to push ahead with five colleges making the change in 1974.

As a college fellow, as well as teaching, one does a fair amount of administrative work, committees, and college offices. I did perhaps more than my fair share as the college dean in charge of discipline during the student revolution of the late 1960s, during which I was also a pro-proctor, a university post which gave me an insight into the higher level of politics. Nonetheless, my main activity as a young academic was overwhelmingly research.

6

PHARMACOLOGY

Sheer luck plays some part in the lives of most successful scientists. In my own case, it has certainly played a huge role. As I have mentioned, I stumbled into being one of the earliest of what one would now term a "computational chemist". I was working on examining phenomena in the area of small gas-phase molecules, in particular their spectra. Then, out of the blue, I received a letter enclosing a scientific paper from Anthony Roe, who was working with Jim Black, later Nobel laureate Sir James Black, for the pharmaceutical company Smith Kline and French in Welwyn Garden City.

Jim, as pharmacologist, had already done monumental work finding the beta-blockers for ICI Pharmaceuticals. Essentially what he had done in that instance was to start with the position that blood pressure is modified by the small transmitter molecule noradrenalin. When this molecule is released, it must bind to a particular protein at a tailor-made binding site. This then causes a change which passes on the message. Nothing was known about the protein or its all-important binding site. Jim's logic was that if one systematically altered the small molecule, one might hit on a compound which would bind into the same site, get stuck there, and thus block the effect when another transmitter arrived at the site. This had worked for beta blockers and had made billions for ICI. Jim then wanted to do the same thing for the small molecule transmitter histamine, which at that time was known to have two effects, labelled as H1 and H2, with two quite separate and distinct physiological effects. Injecting histamine into a person causes firstly eye watering and itching (H1 effects) and separately, acid secretion in the gut (H2 effects). Although antihistamines to block the H1 receptor were widely available, Jim believed that he could find small molecules which would block the H2 receptor protein site and thus potentially cure stomach ulcers. ICI, however, would have none of this and wanted him to stick to beta blockers. This provoked him to resign and join the US company at their relatively small base in Welwyn to pursue the H2 antagonist project.

The paper which I had been sent to comment on was quite simple. Some very crude so-called semi-empirical quantum-mechanical calculations had shown that the small molecule histamine must exist in two alternate shapes, or conformations. The paper went on to postulate that one of these shapes would bind to the H1 receptor protein and the other to the H2 receptor.

I quickly wrote back largely debunking the paper. One could see, without doing calculations, that histamine must exist in one of two possible conformations, often described as "trans" and "gauche", for it is what is known to chemists as a 1:2 di-substituted ethane molecule. On the other hand, I wrote, it was nonetheless a possibility that each of the two shapes of the molecule could bind to a separate receptor binding site.

A very quick response took me to Welwyn to discuss this with Jim and his medicinal chemist colleague, Robin Ganellin. I was quickly appraised of the importance of this problem. They had spent some £12 million following Jim's technique of modifying the histamine molecule and trying to find a variant which would block the H2 receptor. To ascertain whether the conclusion of the published paper had merit, we agreed that the Smith Kline chemists would make a series of compounds of a histamine molecule with one of its hydrogen atoms replaced by a methyl group. I would calculate the population ratio of the two forms of each molecule. They would measure the relative activities, and if there was a correlation between population rates and activity ratio, the question would be resolved.

After both sides had done their respective calculations and measurements, it was quite clear that there was no correlation. However, one of the compounds prepared, 4-methyl histamine, proved to generate excitement since it promoted the H2 activity but not H1. Thus the presence of the simple small group must prevent binding to the H1 receptor.

The Welwyn team continued to modify histamine but kept

the 4-methyl substituent, and this remains as the wonderful compound they discovered, cimetidine, sold as Tagamet, which made many billions of dollars for Smith Kline and French. A substitution in that 4-methyl position is likewise present in ranitidine, made by Glaxo and sold as Zantac, and even more profitable.

This flirtation with pharmacology definitely excited my interest, and I quickly started to set problems in this area to my research students. When discussing with new or potential graduate students, I typically suggested possible research topics. Despite the scepticism of fellow academics, it was the application of our methods to pharmaceutical problems that excited the interest of potential students. So within a couple of years, my entire research effort was in this domain.

The topic was somewhat novel, and I was one of the very few practitioners. In the early 1970s, I became aware that the publisher Butterworths was offering a prize of a couple thousand pounds for an outline of a proposed book. I applied with a suggested outline of a book under the title of 'Quantum Pharmacology'. It was to be in four parts: a brief outline of the essentials of quantum-mechanical computation for pharmacologists, a summary of pharmacology for chemists, a section of one of these applied to the other, and a bibliography of all done in the area up until that time. My outline won the prize. My chief difficulty in writing it would be my relative inexperience in pharmacology.

My solution to that dilemma was to take a year's sabbatical, during which I would have time to write this book and to learn the relevant molecular biology. I wrote to scientists I knew from conferences and at Stanford, Berkeley, and Caltech. I very promptly received an invitation from Oleg Jardetzky at Stanford, who headed the pharmacology department and later from Berkeley, with a one-year commitment to teach graduate quantum mechanics. The two extra salaries proved particularly opportune since, just as I left for California in 1975, the value of the pound sterling fell from $2.40 to $1.80. Furthermore, for just the period when I was in California, the UK Chancellor of the Exchequer, Dennis Healey, altered the rules in that my US earnings were added to my UK salary, and I was taxed at a marginal rate of over 80 per cent. Even so, I think we would have been eligible for food stamps had we been US citizens.

Another slight complication was that just before I arrived in Palo Alto, Oleg Jardetzky fell out with the pharmacology department and had a new lab doing nuclear magnetic resonance. Nonetheless, I was able to learn pharmacology and complete the book.

Another positive aspect of my time in the Bay area was to observe the exciting creation of high-tech companies from university science. I also became acquainted with desktop personal computers. When I got back to Oxford after my sabbatical, I tried to convince my colleagues that PCs were the future but was howled down by sceptics. How quickly things change in the computing world.

This rapid advance of every aspect of computing meant

that we were now capable of doing more and more realistic calculations on pharmacological problems. We could treat bigger molecules and incorporate aspects of the protein-binding sites and the influence of solvents. As well as using quantum-mechanical methods, both ab initio and semi-empirical, it became possible to employ Monte Carlo methods and molecular dynamics.

Nowhere, however, was the advance in computational capabilities more influential than in the use of graphics. Being able to visualise the target binding site and the interacting molecule in a three-dimensional view was fantastic. In my group we produced the first colour molecular graphics images. This was the work of my graduate student Valerie Sackwild, (now Rahmani) and done from a black-and-white screen before colour displays became available. This she did by putting up an image of a molecular structure, photographing it through a red filter, winding back the film, changing the image to that of the electron distribution calculated to surround the molecular structure, and photographing this through a blue filter. Such coloured images had not been seen, and when I showed slides of this work at scientific meetings, I got spontaneous applause. Such was the rate of computational progress that, in no time at all, colour screens became common. The US company Evans and Sutherland, in particular, produced a brilliant commercial machine which permitted the easy use of colour graphics.

My activity in this area came to the notice of IBM. They gave me a machine and instituted a very positive and beneficial collaboration, particularly with Andrew Morffew and Jane Burridge at their Winchester base. Andy very much drove the creation of a new academic society, the Molecular Graphics and Modelling Society (MGMS), the founding group involving me, Frank Blaney of Beechams, Andy Viniter of the Wellcome Foundation, and a few others. The reason for creating the MGMS, apart from organising focussed scientific meetings on the topic, was to create a journal which could publish colour images. At that time, conventional journals charged enormous sums to publish colour pictures. The new house journal, the *Journal of Molecular Graphics,* was initially edited by Andy Morffew, but I soon took over as editor and held that position for many years. I was particularly proud of a brilliantly coloured cover image created by Jane Burridge.

I knew how to edit a journal from an experience I had some years earlier, when I stood in for some weeks as the physical science editor of *Nature,* following a spell with that journal. One thing which was abundantly clear to me was that one needed a top-class secretary who could systematically follow the progress of submitted papers via refereeing, proof, and correcting the publication. I was extraordinarily lucky in hiring Liz Ryder (now Harris), who worked on the journal one day per week. This suited her since she had just had a baby, and the flexible work was convenient for her. Over the next many years, she did more and more for me, finally becoming full time and my personal assistant.

The application of computational methods to drug discovery

gradually became to be recognised as potentially valuable to the pharmacological industry. This was evidenced by the fact that several of the major companies sent one or several of their scientists to spend some time with me to learn about the techniques and often to take versions of our software. Collaboration with industry grew as did the thought of setting up my own company.

Our biggest step forward in the area of computer-aided drug design came with the advent of computer graphics. Being able to visualise protein binding sites was an inestimable help in the design of inhibitors. But one still needed the protein structure. Ideally this came from crystallography, but we also developed methods by which we could predict protein structures. This was much aided by the rapidly growing field of DNA sequencing, enabling us to compare one protein sequence with another. One of the best pieces of work from my group at that time was produced by my then student Garrett Morris. He wrote software to compare sequences and to explore them in detail using colour graphical techniques. Unfortunately, I did Garrett a bad turn by trying to be honest and kind. My contribution to his project was minimal, so when we published his results, I left my name off the paper, leaving him as sole author. As a result, the paper was not as widely read as it might have been had my better-known name been there as second author. However, when we later commercialised our software, his program was a bestseller under the name Cameleon.

Another early entry into a novel field, later to become

important and ultra-fashionable, was that of neural networks, which I started in 1992 with an outstanding student, Sung Sau So, and then taken up in 1993 by one of my best ever students, the Australian Thomas Barlow. Once again, we may have been a bit too far ahead of the game.

7

FUNDING

It may be hard for contemporary academics to believe, but raising funds to finance my research never proved difficult or even time-consuming. I was scarcely able to credit what I learned at Stanford and at Berkeley, where my fellow professors said that they typically spent several months each year applying for grants.

My initial research involved lasers and spectrographs, so I did require funds. But a simple application to the appropriate government body (the Department of Scientific and Industrial Research, the DSIR) was sufficient. Later, all I required was computer time on big national facilities, especially on the Atlas machine at Harwell. When my computational drug discovery started to take off, I applied for a major grant of computer time at Harwell and funds to support a part-time research assistant. This latter aspect was important since there was a lot of tedious inputting of data about each of many thousands of small potential drug molecules, and it did not seem fair to tie up my graduate students doing the routine work rather than producing novel software. I was successful in being granted massive amounts of computer time, but the funding for the assistant was turned down.

This created a dilemma. But by sheer chance, I noticed an advertisement from the Lord Dowding Fund for Humane Research that invited grant applications for research intended to improve human healthcare but did not involve the use of animals. It was not clear how much money they had, so I applied for a box of models to make representations of the structure of molecules and for the research assistant. They rang back immediately, saying essentially that I could have all I asked for as long as I said yes by the following Friday. This caused me to look more closely into the fund which proved to be, as I should have guessed, part of the National Anti-Vivisection Society. I was not sure what to do, so I asked my former Balliol colleague, the distinguished pharmacologist Bill Paton, what I should do. His response was unequivocal: 'They are the enemy; Don't touch them.' Slightly daunted, I also asked my head of department, Rex Richards. His reaction was more palatable: 'Take the grant, but be careful what you sign.' I accepted the grant.They funded me for several years and were impeccable in every way. I did have to give talks for them, often in rather comfortable European surroundings. I always said that I did not believe that it was possible to produce new drugs without animal testing, but that new computational techniques would reduce and ultimately eliminate their use. It was a very happy relationship.

With the funds, I still had to find a suitable research assistant. We did have some contact with researchers in particle physics who were huge users of the Atlas computer facility. My students told me that in the physics lab there were many women referred to as "scanners", who in that era, visually mapped the tracks from particle collision experiments. This rather tedious work was overseen by their boss, Jane Hammond. It occurred to me that my assistant's job would be less tedious, so I asked Jane if she could suggest one of her team who might like to join me. To my surprise, she said she would do my job as she could do alongside her supervising work in physics, where she was a bit like a firefighter, only busy when something went wrong. This seemed a good idea, but not surprisingly, the university baulked at paying two salaries to the same person. Jane, always very sharp and imaginative, then proposed that her boyfriend, a South African graduate student, David Aschman, would take on my assistantship. In fact, the salary was paid to David, but Jane did the work. This lasted until Jane and David went to Stanford, where we met up once more. Jane was succeeded as my research assistant, funded by the Lord Dowding Fund, by Jennifer Wallis who had a PhD in mathematics and was a significant contributor to our work. In turn, she was succeeded by Gaynor Leggate who, with her husband, had run the chemistry department information systems.

My next break in funding once again came to me without my making an application. I was approached by the US non-profit organization the National Foundation for Cancer Research (NFCR) and asked if I would like some funding. The NFCR had an unusual and interesting history which went back to the Nobel-prizewinning Hungarian biochemist Albert Szent-Gyorgy, the discoverer of vitamin C. He also played a prominent role in opposing the Nazis in Budapest during the Second World War. As a consequence, after the war there were moves to make him president of Hungary, but he soon fell out with the communists. Had he not been a world-famous scientist, they would probably have executed him. Instead he was banished and ended up in Woods Hole in America.

He had lost more than one wife to cancer and was determined to research treatments. He was, however, not adept at writing grant proposals and held strong views. In particular, Albert was not impressed by much standard cancer research. He believed that any revolutionary advances were most likely to come from theoreticians. An article about him and his lack of research funding appeared in *Time* magazine, prompting savvy Washington lawyer Franklin Salisbury to set up the NFCR to fund Albert and his work. Being essentially a theoretician, they approached me. However, at the time I could not see how I could do anything useful in the anticancer area and declined their offer, But I said that I would get back to them if things changed.

Not long after this approach, I did see how I would

legitimately do something useful in the anticancer area. The crystal structure of the enzyme dihydrofolate reductase (DHFR) was published. This is a known anticancer target, and with the knowledge of the binding site into which a blocking drug could bind, it was perfect for my sort of research. We could computationally decide which molecules from a huge data base would fit in terms of shape and roughly how tightly they would bind: The stronger the binding energy, the better the drug. The NFCR liked the idea and were very generous sponsors, funding postdocs in my group for many years.

My final novel funding novelty involved little actual work but provided vast amounts of free computer time. This came about by taking the ideas from the so-called SETI (Search for Extraterrestrial Intelligence) project. The background to this project was the fact that on the earth we constantly receive radio signals from outer space and from all directions. There are fairly simple computer codes which can decide whether a given signal is just noise or an intelligent message. The problem is that radio telescopes can record millions, indeed billions of signals, so vast amounts of computer time would be used up trying to detect an improbable signal. A solution to this dilemma was provided by some bright young scientists at Berkeley. The software to detect an intelligent signal was put into a screensaver which was taken up all over the world. The signals to be tested were delivered over the World Wide Web. Computationally, this was very interesting and novel, although ET still has not rung in.

We took and adapted the idea. We produced software to

give a crude estimate as to how strongly a given small molecule would bind into the binding site on a protein known to be a cancer drug target and incorporated this into a screensaver. We then built a database of billions of small molecular structures and sent these out over the Web to people who were using our screensaver. This had more complicated aspects than the original SETI project since for every molecule we sent out, we had to receive the predicted binding energy.

The success of the project amazed us. We reached 3.5 million participants in more than two hundred countries and were given over 450,000 years of free CPU time. The initial work involved Keith Davies, who had founded the Oxford company Chemical Design and was funded by Intel, who financed a Texas company, United Devices, who provided the data transfer aspects. Again we received funding from the NFCR.

Although the overwhelming part of the project was concerned with cancer research, during its course we did take on two other projects and gave the results to the US government. The first of these concerned a search for a drug which would counter the effects of anthrax. Although many people have now forgotten, almost immediately following the 9/11 attacks in the United States, there was an anthrax attack which paralysed Washington merely by putting anthrax spores in the US mail, a far simpler and more effective challenge than a missile attack.

Anthrax toxin consists of three proteins, each of which on its own is harmless. It appears that one of these proteins, the so-called protective antigen, forms a pore to enable the lethal factor to enter the cell. Following on from some work by the group of George Whitesides at Harvard, it appeared to us that there was a site of the protein which, if occupied by a blocking molecule, could prevent the pore from working. That project was funded by Microsoft.

In a similar vein, again with terrorism potential in mind, the world's security services are concerned that if terrorists let out smallpox, they could have a devastating effect. Since the wonderful elimination of smallpox through worldwide vaccination, it might be supposed that it is not a threat. However, some smallpox (variola) was kept for scientific purposes in the United States and the Soviet Union. Following the demise of the latter, there is a lingering fear that some of its material could have gotten into the wrong hands. Because no young people are now vaccinated, any release of smallpox would be devastating. Of course, one could vaccinate the whole population, but this would be against a threat which might never materialise. It would also be expensive and even dangerous for a small but significant minority of recipients. Better would be to have a drug; again, a strategy of blocking a key protein's action is attractive. In this case, the smallpox takes its own very tightly wound DNA into the cell and has an enzyme (topoisomerase) to assist the unwinding so that the DNA can be read. Hence, again we had a protein target to use in the screensaver project. This study, financed by IBM, again resulted in the suggested inhibiting molecule being given to the US government as no pharmaceutical company is interested in a cure for a disease which might never happen.

We finally stopped the screensaver project when it became too successful. The general public really got behind the work, particularly in the cancer aspect since virtually everyone in the world has, or knows of someone, who has suffered from it. I kept receiving heart-rendering emails saying how much someone loved this project: 'My mother has just died from cancer, and by taking part in this project, I feel I am doing something and not just donating money.' They were emails which really demanded an answer, and I was not really equipped to do this when the number of messages grew from hundreds to thousands. This was a major reason for bringing the project to a close. But in addition, the advent of cloud computing meant a much more efficient source of computing power. It became much easier to rent cloud time from Amazon.

The intellectual property behind the screensaver project belonged to the university, and we did set up a company based on the project. My 25 per cent share of the company I gifted to the NFCR, but as will be revealed later, this did not make them wealthy.

8

ADMINISTRATION

Any scientist who conducts his or her research in a university will inevitably have to become involved in some aspects of administration. In Oxford this is more complex and time-consuming than almost anywhere else. This is because three separate organisations require administration: the college, the department, and the central university. As well as creating extra work, complexity is also enhanced by the fact that the three constituencies may, on occasion, be in conflict or at the very least under tension.

When I first became a fellow of Brasenose in 1966, I was immediately appointed junior dean. The job of the college dean is to oversee student discipline. At that time, when many fellows lived in the college, there was a senior dean who was married and lived out, involving himself in only major matters, and a younger resident junior dean who coped with all the day-to-day matters. On taking up the post, I had been assured that there was very little to do, and indeed initially this was true. I was responsible for one early innovation which was the provision of keys to the college gate for the then all-male student body. Difficult to believe, but in that era and until that time, undergraduates who may have fought in the war or other conflicts still had to be in by ten o'clock in the evening or afterwards climb over the college wall. Not only if one was returning late or possibly injured, obtained from playing rugby for the college. Furthermore, women had to be out by 10 p.m. I was prompted to provide keys by Bob Chick, an American undergraduate who had been secretary to California governor Pat Brown (father of Jerry). We were the first college to do so. Bob wanted to have the keys called "chicklets", but that did not carry on. However, his initiative was soon copied across the university.

Although promised that the job of dean was trivial, very soon the worldwide student revolt broke out provoked in the United States by the Vietnam War, but magnified as a general left-wing uprising across the world, notably in Paris, and even troubling Oxford. This did involve me, not just at the college level, but more widely since, at that precise time, I became a pro-proctor. The post of proctor is a very ancient Oxford office. Each college appoints one of two proctors who serve for one year. The proctor sits on all university committees, controls examinations, and plays a visible role in ceremonies. It is often a way of entering university administration and politics.

Each proctor also appoints two of his colleagues to act as pro-proctors, who take on some of the more tedious and lesser roles. I was a pro-proctor in the during the period 1969–1970, the turbulent years. When I did the job, the proctor, or more often we pro-proctors, had to walk the streets in the evening wearing full academic regalia and supported by the university marshal and two of the university policemen, wearing bowler hats and known as Bulldogs. If I saw any misbehaviour, I was supposed to get one of the Bulldogs to approach the young man, raise his hat, and ask, 'Are you a member of the university, sir?' By the time of the student revolution, the usual response was, "F*ck off!' Hence that custom died out in my time. More irritatingly, I had to spend a night in the university offices awaiting the leaked break-in by the revolting students. Their ineptitude meant that they broke into the offices of the Wytham Woods Rabbit Clearance Group by mistake, so the protest was a flop. I did sit through some of the trials of the main student activists, but by international standards, we were relatively calm.

One life-changing result of being dean came towards the end of the student revolt. One of their leaders, later the distinguished High Court Judge Michael Burton, organised a party in Trinity College to which were invited the student leaders, the presidents of all the college junior common rooms, the deans, and the proctors. I went along with the not admirable intention of picking up one of their girls. In fact, I met the president of the Somerville College Junior Common Room, was immediately attracted, and in 1970, Jessamy and I were married. We had eighteen wonderful years until she died of cancer in 1988.

Later in my Brasenose career, I became senior tutor, a job involving the organisation of the tuition of the undergraduates. These days this is a full-time occupation performed by a specially appointed fellow. I did this job in the late 1980s and early 1990s. It took me about an hour a day, with most of the work and all the bureaucracy being done by the wonderful college secretary, Wendy Williams. As at other times in my career, I was totally dependent on an intelligent personal assistant, often a woman.

Apart from these formal jobs, I sat on college committees, most notably for many years on the Finance Committee. In that respect, one innovation for which I can claim credit is the living-out allowance paid to fellows who did not reside in college. When I was first a fellow, some dozen of the twenty-odd fellows were bachelors and lived in rooms in the College. Married fellows either lived-rent free in a college house or were paid a "marriage allowance". This is a contentious and provocative business, still extant to some extent. Since the rewards vary from college to college, major discrepancies and unfairness can occur. Brasenose, during my time, got rid of college houses and invested the money on the London Stock Exchange rather beneficially. Before I married I bought a flat and moved out. I was obviously not eligible for the marriage allowance, although divorced fellows still drew the money. I managed to get the system changed to being a living-out allowance.

My final college post was vice principal, a job passed down

in order of seniority among the senior fellows. In my time, the chief role was to take the minutes at college meetings, an activity now done by a professional secretary.

When I first joined the physical chemistry laboratory, the entire administration was done by the head of department's secretary, the formidable Miss Binnie, nowadays replaced by a cohort of more than thirty. There are many stories about Miss Binnie. My favourite concerns the time when a boring bureaucratic government body demanded that all laboratories should produce an inventory of all bits of apparatus costing more than £500. It would have been an enormous task. Miss Binnie merely went to the filing cabinet, removed the label saying "Invoices", turned it over, and re-labelled it, "Inventory".

Rex Richards was a joy in terms of administration. Although he really made all decisions, he held a tea party in his office at 4 p.m. every Monday, where we agreed to do what he wanted to do anyway. There was quite a group of us newly appointed lecturers who started in the mid-1960s. We developed a sort of ritual whereby we would meet at lunchtime on Mondays in the back bar of The Mitre Inn and come to a joint view on matters of departmental politics. This grew more formally into a club, the Hinshelwood Club, which still exists. One lasting legacy of the group derives from the fact that on his death, Hinshelwood left his estate to the Goldsmiths Company in London. Inexplicably, that wealthy organisation sold his Nobel Prize medal on the open market without even consulting the department or the university. Incensed, we embarrassed them into funding an annual Hinshelwood Visiting Lecturer. (The university would not let us use the description "Visiting Professor".) That post still exists, although not financed by the Goldsmiths, and has had some stellar lecturers from all over the world.

Until I took over as head of all chemistry in 1996, the subject boasted three separate departments, one each for the three tribes: physical, organic, and inorganic sub-disciplines. The overseeing body was the chemistry sub-faculty, with representations from each branch and an overall chairman and secretary. I became secretary to the chairman Brian (later Sir Brian) Fender. This was a real and valuable education. He was a consummate politician as the French and Germans found when he later became head of the Institut von Laue Langevin in Grenoble. I well remember a contentious meeting when Brian spoke against the majority who supported a particularly contentious move. He lost. Afterwards, I said to him, 'But surely that was something you really wanted, yet you voted against it.'

'Ah,' he replied, 'that is true. But if I had voted for, X and Y would have opposed.' Thus he got his way and went away chuckling.

My first significant role at the university was as chairman of the University and Industry Committee, a device set up during Harold Wilson's "white hot technological revolution". We had carte blanche and did a number of novel things. For instance, we included outsiders on the committee, including Richard Charkin from the University Press and David Young of Oxford Analytica. It is fair to say that the central university administration did not approve of us, so after I ceased to be chairman, they quietly abolished the committee.

More significantly, I was elected to serve on the general board of the university, probably the most important committee in the university, only slightly junior to the top Hebdomadal Council (so named as it met once every two weeks). The general board ran all things academic in Oxford. While serving, the North Report came out, and the general board and Hebdomadal Council were abolished and replaced by a single council. This was filled by a university-wide election, as a result of which I became a founding member. Here too there were outsiders in the sense of not holding a university post. Most valuable were Sir Victor Blanc and Bernard Taylor, who brought some economic sanity and proved very influential. Later, when I had to raise the money to fund the new chemistry laboratory (see chapter 11), my presence on council was enormously valuable, not least in my knowing how to work the university system.

9

PUBLISHING

For all scientific researchers, publication of papers in high-quality journals is imperative. The old tag, 'Publish or Perish', is very true. Throughout my career I have had a research group typically of around a dozen people made up of chemistry part II students, doctoral students, and a couple of postdocs. Thus, over most of my career, I published ten or a dozen papers each year. A full list is given in the appendix.

I have also published some twenty books, again listed in the appendix. This aspect of my life started very early. My research involving the use of ab initio molecular orbital quantum-mechanical calculations was in an area more generally considered as physics rather than chemistry. To encourage chemists to venture into this area, I and my first doctoral student, John Horsley, wrote a book aimed at chemists. This habit of writing books with my students as co-authors became my standard practice.

Nowhere was this more effective than when I took to writing textbooks for students. For several of them, my collaborator was my student Peter Scott. We had a very successful partnership. I seemed able to dash off a quick first version, perhaps containing errors. Peter, by contrast, was meticulous and painstaking. I must emphasise he was not just a proof corrector. When he looked at the text, he instituted major changes, including new chapters. It was a very happy relationship with close to fifty-fifty input.

There were a couple of reasonably influential texts which I did write without a co-author. My book *Quantum Pharmacology* opened the eyes of a significant number of theoretical chemists to the possibility of applying their techniques to problems in biology and indeed to drug discovery. That book was translated into five languages and had a second edition. It also helped build my relationship with the pharmaceutical industry, which in the early days sent young members of their research departments to spend some time in my lab and to take some of our software back to their companies. That relationship also, in part, provoked my creation of a new spin-out company, Oxford Molecular, and in turn my books about the founding of companies based on university intellectual property.

I gained further insight into scientific publishing by spending some time with the prestigious journal *Nature*. *Nature* was transformed into one of the two higher-rated scientific journals, rivalling the US journal *Science* when John Maddox was its editor. Amongst other coups, he very rapidly accepted and published the Watson and Crick paper on DNA. Despite his brilliance, he was, however, a somewhat chaotic editor. The publisher, Macmillan, moved him upwards, and the editorship passed to David Davies, who was much more disciplined. He introduced a system called Nature Sabbatical Writerships. I was one of its first holders and spent a very informative month with them. During that time I wrote a number of 'News and Views' pieces. The following year, the physical sciences editor intended to take time off to attend some international scientific meetings. I stood in for a month and really learned how to edit a scientific journal. Particularly with *Nature*, many papers came in each day, but not all could be refereed. That first cut was crucial, with those not selected prompting a standard letter suggesting "another journal". Then one had to pick referees and judge their responses. This was before the era of modern computational systems, and there were large piles of paper to shuffle about. This experience was vital when I became the editor of the *Journal of Molecular Graphics*.

One of my most interesting experiences in publishing came when I became chairman of the publishing division of the Royal Society of Chemistry, on whose council I sat and was at one time vice president. At that time, before the advent of open access, academic publishing was a very profitable business. Indeed, from an annual turnover of more than £30 million, the Society made a profit of approaching £20 million, that financed most of its activities. When I took charge, as I had done in other areas, I brought in experienced outsiders, notable Chris Lowe, Peter Doyle, and my former student Reg Hinkley.

One activity being developed at that time was the creation of new journals. Initially I was somewhat sceptical about this but soon learned just how successful and profitable this could be.

While doing this fascinating work, the OUP, without consulting me despite the fact that I was at the time head of Oxford chemistry, decided to cease publishing chemistry monographs. The Royal Society of Chemistry publishing took this up and made a million-pound profit in the first year. This was not my only skirmish with the OUP. My first significant problem occurred when I went to California to teach the summer quarter at Stanford. I sent details of one of my OUP books that I wanted the students to use. When I arrived, I was irritated to find that the campus bookstore did not have it in stock. On enquiring I discovered that OUP New York had informed them that it was out of print, which was untrue. The New York branch operated totally independently and, for instance, did not publish for the United States market the spectacularly successful book of my brilliant author colleague Peter Atkins. These very successful OUP books were published in the United States by Freeman, who made a huge profit.

I was fobbed off when I got back to Oxford by being made a consultant to the OUP. But this did give me an insight into what I saw as massive inefficiencies and stupid moves made by the institution which was, and is, just a department of the university (very tax advantageous). For instance, they set up a paper mill with the Britton family, to whom they lent the money to fund this venture which totally failed. This provoked me to propose that the university should float the press on the stock market and retain just 51 per cent. I wrote about the suggestion to the vice chancellor, who did not respond. I then published my idea in the *Oxford Magazine*, and the suggestion received national publication. I received heavy rebukes from the vice chancellor, who even swore at me. Although the university missed out on billions by not taking up my idea, my campaign did have some effect. Profit made by the OUP used to stay within the press, who occasionally made generous gifts to the university of our own money. Following the rumpus, the press subsequently, in a much more formal way, subvents the university. However, the danger of holding a huge portion of one's assets in a single investment remains dangerous.

10

COMMERCIALISATION

In 1978 I became chairman of Oxford's University and Industry Committee. As mentioned previously, this committee had been set up as a consequence of the Wilson government and its "white hot technological revolution". All universities were supposed to have such a committee, although they were not too sure what they were meant to do. Vice chancellor Alan Bullock appointed a personal friend, a former headmaster, to run the committee with mathematician Alan Tayler as its first chairman and a rather high-powered membership, reporting directly to the Hebdomadal Council and general board. When I succeeded to the chairmanship, Michael Day of the Careers Service (then called the Appointments Committee) became secretary.

We had an essentially clean canvas and could do almost what we wanted to do, although we were used as a dump to which daft government papers could be sent to be looked at and then commented on. As far as industry was concerned, that seemed a very distant and foreign world. Some inkling of just how different can be gleaned from the fact that in that era, the attitude of the central university was dominated by the notion of risk. It was even the case that a letter was circulated to all the scientific faculty stating that it was not permitted to do consultancy on university notepaper. The university was afraid of being sued.

This fear of litigation was not without foundation. It dated back to the 1920s, when the university had a professor of agricultural engineering who invented a machine to extract sugar from beets. It was licensed to an Italian food company. But unfortunately, Professor Owen was a swindler who reputedly ended up in Dartmoor Prison, where he even swindled the warders, and the university was sued for a not inconsiderable sum. The first useful thing my committee did with the help of the university's finance director (then called the secretary of the chest) was to take out insurance to cover the university when its employees gave advice and acted as consultants.

My own interactions with industry were much influenced by changes wrought by Mrs Thatcher during her tenure as the first scientifically trained premier. One of her first moves was to introduce venture capital (VC) into the United Kingdom by changing the tax regime to encourage the backing of high-technology ventures. One of the earliest groups to take advantage of the new situation was the US chemical company Monsanto, who had had a venture capital fund in the United States for many years. As well as producing a return of more than 20 per cent per year, this fund was a vehicle to enable the company to keep an eye on novel technologies and, if appropriate, bring them in-house. Monsanto appointed DeLoittes as consultants to help them set up a VC fund in London, but with the added part to the brief being that as a result of their US experience, they wanted their UK fund to be linked to top universities.

The DeLoittes' consultant was an old school friend of mine, the late John Winter, who rang me to ask to whom he should talk in Oxford. Given that I was chairing the University and Industry Committee, he had the appropriate contact immediately. Along with the Brasenose bursar, Norman Leyland, who had founded the Oxford Centre for Management Studies, later Templeton College, we organised a meeting of college bursars, several of whom subscribed to the fund.

John Winter also found the appropriate manager for the fund, which became Advent Eurofund, in the person of David Cooksey, now Sir David, to invest and manage the fund. Monsanto also had one of its own employees seconded to the fund management, Paul Bailey, an Australian whose father had been one of the earliest Rhodes Scholars.

David Cooksey set up an advisory committee consisting of Sir Peter Hirsch, the Oxford materials professor; Sir Hans Kornberg, the Cambridge biochemistry professor and master of Christ's; Bruce Sayers, the Imperial College computer scientist; and me. One of the things this group was expected to do was to spot likely new technologies and possible investments. Eager to earn my fee, that represented a significant addition to my income and permitted me to send my sons to The Dragon School, I directed the attention of the fund to the work of my inventive colleague Raymond Dwek in the biochemistry department. He was doing some lovely original work in a field he christened "glycobiology".

It was decided that the technology was a bit new for setting up a company, but Monsanto was sufficiently impressed for it to become major funders of Raymond's work and even to sponsor a building. I did thus earn my fee. And a spin-out company, Oxford Glycosciences, did also later come to fruition.

Mrs Thatcher's second vital innovation had an even greater impact. In 1987 she caused the ownership of the intellectual property of research conducted in universities with government funding to be transferred to the universities, providing they set up mechanisms to exploit the intellectual property. The background to this seismic change is not as well known as it ought to be. The history goes back to the Land Lease Agreement signed by Roosevelt and Churchill in 1940, when the United States gave Britain fifty ships in return for leases on bases in the West Indies. What is less well known is that in the small print of the agreement, the United Kingdom agreed not to patent radar, the jet engine, and penicillin. The wartime reasons for this are obvious, but when peace was restored, it was clear to the Atlee government that hundreds of billions of pounds had been given away.

The Labour Government set up the National Research

for Development Corporation (NRDC), later the British Technology Group, to exploit all the intellectual property that developed from state-funded research. Although large returns were achieved from the pyrethroid herbicides discovered at the Rothamsted Laboratories and from the cephalosporin antibiotics originating from Oxford's pathology department, they chose not to exploit either Cockerell's hovercraft or the monoclonal antibody work of Cesar Milstein. This error provoked Mrs Thatcher, stimulated by David Cooksey, to make the change giving ownership of such intellectual property to the universities. This was to have a profound impact.

In Oxford, we had advance warning of this change. Again by pure chance—and I am constantly repeating that I am a firm believer in the "Cock up" theory of history—I was hosting at a meeting of the Natural Science Club with David Cooksey as my guest. He was speaking about venture capital and the new situation with respect to intellectual property. David suggested that he could come up with the funds to set up a technology transfer company or organisation. The following day, Peter Hirsch and I wrote to the registrar, urging this to be followed up. It was, albeit very slowly and even grudgingly. But in 1988, the university set up Isis Innovation Ltd (now Oxford University Innovation Ltd), a wholly owned subsidiary whose remit is to exploit the university intellectual property by means of licensing or the formation of spin-out companies.

I was on this original management committee and then became a director of Isis for the next twenty years. I did play a part in the creation of Isis Innovation, but its implementation and success were due to the first CEO of Isis, James Hiddleston. He set up Isis Innovation Society, whereby for an annual fee companies can attend three dinners with networking opportunities and a thirty-day exclusive view of licensing opportunities. This method I had somewhat earlier used to raise £600,000 for the Oxford Interdisciplinary Research Centre, the Oxford Centre for Molecular Sciences, which had grown from the Oxford Enzyme Group.

At an early meeting of the Isis board, when we were looking for some examples with which to start things moving, I suggested the formation of a company based on my research in the area of computer-aided drug design. This was a notion which had been in my mind for some years, ever since I found that pharmaceutical companies were asking for copies of the software written by my graduate students. Since the early 1980s, I had either given copies of the computer programs free of charge or for a nominal sum since the software was not written to professional standards. And moreover, I was in no position to support anything we actually sold. There was no doubt that we possessed some technology which could form the basis of a company, but even at that stage, I realised that even more important than the science was the person or people to run any venture.

Happily, there did exist the ideal man. In the late 1970s I had as an undergraduate pupil at Brasenose Tony Marchington, who was a man in a million when it came to commercialisation.

The son of a north Derbyshire farmer, Tony was a natural trader. My first inkling of this was during his research in my group when he earned a few thousand pounds by part-writing the script for a film about C. S. Lewis, an opportunity which had arisen because Tony happened to be in the same digs as Walter Hooper, who had been Lewis's private secretary. It was what Tony did with the money which marked him out as an entrepreneur: He bought a steamroller! At first I thought this was crazy, but in fact he took the roller to steam rallies and was handsomely paid. So it was profitable as well as being his hobby.

For his D.Phil, Tony became one of the first students to be supported by a so-called CASE award, a cooperative award for science and industry. This had to be held in conjunction with a commercial company. In Tony's case, it was the ICI Plant Protection Division, their agrochemical business, now part of Syngenta. On completion of his doctorate, the company was very keen to employ him and to take my technology into the company, and they made him a very generous offer, despite this being a period of downturn with a company-wide no hiring policy. On the basis of his letter of employment, Tony borrowed £35,000 from the bank and bought a pair of steam ploughs, huge steam engines which work by pulling the ploughshare between them. These, too, he took to steam rallies but very quickly formed his own company to run rallies. This was initially a sideline, but in the mid-1980s, he left ICI and just ran steam events.

The final fact which caused me to try to set up a company was the death of my wife, Jessamy, from cancer in November 1988. The day following her funeral, I rang Tony and suggested that we do something about the company which we had discussed over the years based on the computer-aided molecular design research in my group. The tipping point was my need to keep busy as a form of therapy after the loss of my wife. Tony was keen, and since his steam engine activities were concentrated in the summer months, we had the winter to get things set up.

Aided by James Hiddleston and Isis Innovation, we produced a business plan with which Tony could try to raise the necessary finance. This was not easy as the VC world was not warm towards software companies. Our friends at Advent turned us down. But Paul Bailey had by that time left them to join the VC arm of Barings, then called Baring Brothers Hambrecht and Quist, came up with some cash, the tail end of one of his funds. The other original investors were the US fund AMT of Delaware in the shape of Peter Walmsley, and most importantly, private money from Roderick Hall, who became our staunchest backer and ultimately chairman when an IPO loomed.

The company was started in November 1988 with just £350,000 of venture capital and the equity being split in equal thirds to the VCs; the university; and the founders, who included Tony Marchington, Tony Rees of biochemistry, who also contributed some software, and me. This was enough money to run the company for about six months. We rented premises in the university science area, the old workshops of astrophysics, a terrapin hut just by the parks. We hired two employees: a technical officer, another former pupil of mine, David Ricketts, and a secretary. Tidying up our software was done by contract programmers from Tessella.

At the end of the first exciting six months we did have products to sell, but income came rather from special deals achieved by Tony. He fixed up porting deals with Hewlett-Packard, who needed software to demonstrate their new computer workstations to the pharmaceutical industry. And even more innovatively, he made arrangements with Glaxo, British Biotechnology, and later with Roussel and SmithKline Beecham, whereby they paid us up front for software yet to be written and the promise of future products free of charge, although not free of support and maintenance. We also introduced a novel scheme to provide our software very cheaply to academic researchers using a government scheme run from the University of Bath. The income was small, but relationships with universities were important.

Right from the start we intended the company to become an international player and early on did a deal with the Japanese company Toray. We entertained them in the Brasenose Senior Common Room, complete with candelabra and especially snuff being an excellent public relations boost.

After eighteen months we were marginally profitable and had convinced our VC backers that we had a future and could run a company. They agreed to provide second-round funding at a price equivalent to double the original share price. So we proceeded to open new offices in the first building on the new Oxford Science Park, owned by Magdalen College. We also opened a sales office in Palo Alto. This was done in the summer of 1991. But the night before the September board meeting, at which the arrangements were to be formalised, our VC backers rang to say that since we had not made as many sales over the summer as we had anticipated, they would still provide the funding but at the original share price rather than at a multiple. In effect they were shafting us, but Tony had anticipated such a stunt, and we managed to bluff them that we did not need their cash as we had an alternative backer. They were unsure enough to drop the idea and did provide the cash at the previously agreed price. But it taught us a lesson about the priorities of VCs. They, understandably, aim to please their own shareholders and are not friends to the companies they support except in so far as benefits are at least mutual.

With some of the money derived from the second-round funding, the company did its first takeover. The target was a French company, Biostructure, based in Strasbourg. They were effectively in receivership so that the transaction had to be with a French judge. Once again Tony played a blinder, hinting that he was losing interest in the purchase by gazing at his airline ticket while the judge was pontificating. This so rattled the official that he dropped the price by 50 per cent. One result of the purchase was that Remy-Cointreau became one of our biggest shareholders. After the transaction I entertained their chairman, M. Herve de Broeill, to a dinner in Brasenose and laid on our very best sauternes. His reaction was wonderful: 'Ah, my cousin made that wine.'

By the summer of 1993, the company could be said to be thriving. This coincided well with one of the periodic windows of opportunity to float companies on the London Stock Exchange, or as the current jargon has it, have an initial public offering or IPO. A number of city groups approached us to see if they could take us public, including some prestigious names, but we chose to make the attempt using Henry Cooke Corporate Finance of Manchester, already familiar as one of the biggest provincial stockbrokers but also connected with Brasenose through the Hulme Trust, which they managed to the benefit of both the college and the university.

For the IPO, Rod Hall took over from me as chairman, and at the insistence of the Stock Exchange, we brought in a new non-executive director, Christopher Weston, the chairman and CEO of auctioneers Phillips. We also had to engage a new managing director and a finance director.

When the prospectus for the float was ready and agreed by our lawyers, there had to be a "road show", where we explained why we needed and deserved funds to city institutions such as pension funds. The twenty-minute show had three parts: Tony introduced the company, I or Tony Rees explained the science, and then Tony covered the commercial aspects. In the rehearsal I was far from convinced as Tony spent most of the opening part talking about Isambard Kingdom Brunel designing and building the railway bridge across the Thames at Maidenhead.

When it came to the real thing, however, this part went down very well, while my science caused eyes to glaze over. We were successful, selling one third of the company for £10 million, but it became clear to me that the chief criterion for those who chose to support the float was confidence in the people, particularly Tony, rather than the technology.

The funds provided by the IPO enabled the company to pursue ambitious goals. Above all, we went in for a series of takeovers in the United States, starting with the Palo Alto company Intelligenetics, bought from the oil company Aramco. Their purchase was made with paper—that is, with shares in the company—and was similarly followed by the acquisition of CAChe Scientific from a Sony-Tektronix joint venture; MacVector from Eastman Kodak; and a computer-aided toxicology business, Health Designs Inc, based in Rochester, New York, where the leading scientist was yet another former postdoctoral researcher from my group, Vijay Gombar.

By this time, we had outgrown Henry Cooke as our advisors and switched to Barings as our bankers and Cazenove as our brokers. Having these powerful groups behind us was particularly important since, following the 1994 flotation, all the original shareholders, the VCs, the university, and the founders were locked in for two years and not permitted to sell shares until 1996. Perfectly legitimately all three parties hoped to exit to some extent to raise cash. Thus, once again, a road show was essential to encourage Cazenove's clients to buy shares. Given that 60 per cent of the share might in principle be up for sale, one might have expected a slump in the share price. But thanks to the skill of the broker and our performance in the road show, the share price just moved steadily onward and upward. The university sold £10 million of its stake, and the VCs exited with massive profits. The only mild distaste was that Tony and I had been told that we could sell half our shares, but the night before the due date, he and I were rung up by Cazenove and told that the City would not wear our selling so much, and that we would be restricted to getting rid of only 25 per cent of our shares. For me this was irritating. For Tony it was more serious as he had already spent the money buying a farm, and more spectacularly, the steam locomotive Flying Scotsman and its carriages.

For a time the company thrived. We were, for two consecutive years, the second-highest rising share on the London Stock Exchange, with the value of the company reaching as much as £450 million. We were able to fulfil early dreams, going beyond being a software company which was in the area of designing drugs to add subsidiaries which could make the compounds using the then-fashionable technique of combinational chemistry, and also another subsidiary capable of testing the compounds. These three arms could act independently or in any combination to do deals with major pharmaceutical companies. Our most successful such venture was with the Japanese pharmaceutical company Yamanouchi in a £10 million project.

We also made further acquisitions, including buying GCG Wisconsin, making us the world's biggest players in the fashionable area of bioinformatics, with 25 per cent of the world market and 60 per cent of all sales in the United States, a rare case of a UK high-technology company succeeding over there. Indeed, half of the four hundred employees were located in the States.

From this high point, however, things started to unravel. Perhaps we had grown too fast. Perhaps Tony, who had been the genius behind the start-up, was not the right CEO for a biggish company spread across the globe, although our advisors did not want to see him go. Certainly we felt that our brokers were selling us short. As soon as the share price started to fall, things quickly became ugly, with a real stand-off between Tony and the board. In the end, after I resigned as a director, the company was sold in two parts, both to American companies for some £70 million.

Looked at from where we started, that seems pretty satisfactory. But viewed from a market capitalization of £450 million, it is disastrous rather than just sad. It was, however, at the very least an education for me and encouraged me to continue in this world.

The success of Oxford Molecular probably made me too cocky. My research group was very much into the use of the internet from the time it was ARPANET. As a consequence, we were involved with the use of the World Wide Web from its inception. Clearly it offered amazing possibilities. In my research notebook, I wrote down three ideas of things we could do on the Web: an internet auction house (later appearing as eBay); an internet bookmaker (now multiple examples), and an internet conference business. It was the latter choice we made with my then postdoc Barry Hardy as CEO and funding from me and my entrepreneurial and generous friend Roger Bilboul. We called the company VEI Ltd, and trial runs of conferences which were laid on for free were successful. However, once we started charging, it failed and had to be wound up. We had made the classic British errors with novel technology: going in too soon and with insufficient financial backing. Nonetheless, I still think the idea is worth pursuing. It would cut down the enormous time senior academic scientists spend in the air, going to conferences, saving both time and greenhouse gas. It was not purely a joke when I used to say that if I wanted to meet up with certain of the prestigious Oxford colleagues, the best chance was to be at Heathrow.

My most recent company formation, which still runs on well into my retirement, is Oxford Drug Design Ltd (formerly Inhibox Ltd). This business had its origins in my screensaver project, mentioned in a previous chapter. As a result of that project, we have a database of billions of small drug-like molecules, all of which we know how to synthesise, and software to estimate the binding energy of a small molecule to the binding site of a protein known to be a drug target.

My former research pupil, Paul Finn, became the CEIO, and we started life as a fee-for-service company. There were some deals with major pharmaceutical companies and biotechs. We had a particularly good joint venture with the brilliant Irish entrepreneur Lord Ballyedmond. He understood well how to take molecules already patented and tested for safety and to find novel targets for these, particularly in veterinary products. Things were going very well with him until tragically, he was killed in a helicopter crash.

Similarly, we had a project with Galderma, part of the L'Oreal group, aimed at finding compounds to treat psoriasis and skin complaints. We produced some novel compounds which really excited the Galderma scientists, but higher up, their company directors feared that our compounds would compete with their existing products. So they just sat on our work with the result that our anticipated royalties were not to be forthcoming.

The company was about to go under, but fortunately Paul Finn is a genius writer of grant proposals. He was successful at raising major grants from the European Union, who had put out a call for funding requests in the area of antibiotics which are so desperately needed but not favoured either by the pharmaceutical industry, which prefers drugs that one needs to take for several years rather than curing a condition in a couple of weeks, or with the financial world. Just as these grants were running to the end of their time, our major shareholder, IP Group (of which I had been involved in founding and of which I was at one time chairman), refused to put in more cash. Yet again, Paul's grant proposal skills saved the day, and we received some £10 million in funding from the US-based CARB-X fund and from the Department of Health and Social Care.

We now have some very promising, quite novel antibiotic

leads which are active against the difficult gram-negative bacterial targets.

The company remains small with all its in-house research done on computers. The designed molecules are synthesised either in Latvia or in India. Testing is also outsourced.

It is sheer delight to work in a small high-tech company. In many ways it is very similar to working with and in an academic group. Given my quite wide experience in the start-up game, I have also had the pleasure of serving as a non-executive director of several Oxford spin-outs, being chairman of a few and on the scientific advisory boards of others. The whole scene has developed extraordinarily over the past thirty years or so. The commercialisation of academic research has become a vital part of university life. This is largely about money, but not totally so. Clearly, if successful it generates cash for the university and indirectly for the government. In addition, it creates jobs for people with doctorates and for postdocs, and it keeps academics in universities rather than having them leave in pursuit of jobs outside academia.

There could be a danger that this type of activity would over-influence the choice of research topics. This is not so. The most successful spin-outs have come from blue skies research, not from the obviously applied. I always like to recall the wise words of the former Royal Society president George Porter, who said that there are only two sorts of research: applied research and yet to be applied research.

11

NEW LABORATORY

Every decade or so, the university conducts a review of major departments with a high-powered committee, including overseas scholars, often our Cambridge equivalents, and senior Oxford colleagues. The review of chemistry in 1986 came up with two major recommendations: There should be an overall head of chemistry rather than having three separate departments for the three historic subdivisions of the subject; and further, that there was a desperate need of new laboratories to replace those existing and barely meeting the norms of health and safety.

When a similar review took place in 1996, the identical conclusions were reached. Plus the rider asking why nothing had been done for the previous ten years. The answer to the latter question was simple: The three existing heads, all big men, were happy for there to be one overall head as long as it was neither of the other two. Thus in 1996, it was resolved that there should be an amalgamation to create a unified department with a single overall chairman. Plus, this person should not be one of the existing heads, and it should be an internal appointment. An election was held, and I ended up in the post.

To be clear that I was as far as possible non-partisan between the three tribes, I had my office set up in the old pharmacology laboratory which had been renamed "New Chemistry", later "Central Chemistry", and now the site of the new Earth Sciences Laboratory, in which all three branches were represented. As my administrator I chose Sarah Hargreaves, who had run the cross-departmental Interdisciplinary Research Centre (IRC) in molecular sciences, essentially proteins, the successor to the very successful Oxford enzyme group. I kept my long-serving secretary, Liz Ryder (later Liz Harris), since one of the most important lessons I had learned after years in academia was that successful administration demands having a first-class secretary or personal assistant with whom one can be totally open and frank.

On taking up the post in 1997, I saw three main tasks in my in-tray. These were firstly to reorganize and upgrade our IT. At that time we did not have standard email addresses, and many colleagues were not competent to handle information electronically. By appointing Karl Harrison to oversee our efforts in this area, we soon led the university in this respect and had a departmental website which received far more hits than any other. The second challenge was to revise and upgrade the undergraduate course which I felt, apart from the wonderful part II year of research, was awfully old-fashioned. In this I was almost totally unsuccessful. Too many colleagues were not prepared to change their teaching.

The final massive task was to provide new research facilities. Prior to becoming vice chancellor, Colin Lucas chaired a committee to look at future needs and sites for new developments, and chemistry lobbied successfully to be allocated what most Oxford people referred to as the "Oxfam Car Park", the car park on the corner of South Parks and Mansfield Roads, run by the charity on Saturdays. We preferred this site to the alternatives of refurbishing existing laboratories or going to a green field site outside the city, for example at Harwell. I had seen how non-productive laboratories are when everyone leaves at five in the afternoon. In a university lab, one needs to be close to the centre of town so that graduate students can have access twenty-four hours of the day and can, for example, come back after dinner to check on their experiments. We hired the architectural practice RMJM to conduct a survey of what we needed and rough costings. It was clear that on the site we could build a building which could house all the organic chemistry research, then occupying the historic but smelly and barely safe Dyson Perrins building about half of inorganic chemistry and perhaps a quarter of physical chemistry. So far this was essentially hypothetical with a notional cost of about £60 million.

These notions would probably have remained just that had it not been for the timely introduction of the Joint Infrastructure

Fund (JIF) with cash from the Research Councils and The Wellcome Trust. It was clear that we should make an application as our case was strong, but it was not clear as to just how much money we dared ask for. My soundings in the central administration suggested that the maximum we could request was £30 million. However, as a department we had a very effective advisory council which included in its membership both Dame Bridget Ogilvie, the former director of The Wellcome Trust, and Peter Doyle, the ex-AstraZeneca research director and chairman of the BBSRC research council. They counselled asking for the full £60 million. This caused some consternation with the Research Councils, but we persisted and were awarded £30 million, the biggest JIF grant given to anyone. Had we asked for £30 million, we would probably have received £15 million. With this sum in the bag, providing we got planning permission, the project would become real. RMJM drew up detailed plans with not a little help from major pharmaceutical companies, which had themselves recently built new laboratories, notably Glaxo Wellcome, SmithKline Beecham, and AstraZeneca.

Our first scare in respect to the site was that we would lose it to the Said Business School, which had just had its plans to build on the Merton Field rejected. Happily for us, Wafic Said did not like the site and preferred to locate his building by the railway station.

Getting the plans past the city planners was far from straightforward, but happily the Labour chair of the committee was Maureen Christian, the widow of materials professor Jack

Christian and a strong proponent of modern architecture. She was an old political hand and knew all the tricks. She brought up the issue at the planning committee meeting just before the lunch break, when she knew folk were keen to get away. We succeeded by one vote. Unfortunately, the rules of the game are that members of the committee are able to demand a vote of the full council, and this was done by the wretched Greens, who opposed any development. The vote at the full council was crucial, and we would have lost over £30 million were it to have gone against us. Watching from the public gallery, things did not look good. The council had a very slim Labour majority with a lot of Lib Dems and Greens who were temperamentally opposed to a new laboratory. Once again, Maureen Christian's knowledge of how to play local government politics saved us. She demanded a named vote. Many of the Lib Dems were actually university employees and could not comfortably oppose, and so they abstained. She personally telephoned each Labour councillor with individual arguments as to why the building was a good thing, be it employment opportunities or aesthetics. Again we won by a single vote. When we came to start the construction, it was Maureen who broke the ground with a JCB. We now had a massive start towards the funding of the project to which another sum approaching £10 million was furnished by a HEFCE scheme. This still left a shortfall of nearly £25 million when cost escalations were included.

Charities were our next avenue for support. The E. P. Abraham Trust, with funds derived from the work in the Dunn

School of Pathology that yielded the cephalosporin antibiotics, contributed half a million pounds. We were also successful in raising a large sum from the Wolfson Foundation. This was not straightforward, but it was very rewarding. Lord Wolfson took a very close personal interest in how his charitable donations were employed. We visited him and his advisors, taking with us the architects and a model of the building: He would have preferred brick. He also made it quite clear that he would not pay VAT and that his limit was £2 million. At the same time, he was clearly very taken with my ideas of a seminar room which would be equipped to webcast lectures. This emboldened me to ask for £2 million to furnish one floor of the building, plus £1.5 million for the seminar room. Wonderfully, he gave us the full £3.5 million.

I had hoped that British industry, particularly the pharmaceutical industry, would be major contributors. In fact, they provided nothing, not even my old friend Tom McKillop, who I knew from Parisian postdoc days and then CEO of AstraZeneca. The only money we received from British industry was £250,000 from Thomas Swan. Tom was a Brasenose chemistry friend. In this case, because his successful and innovative company is privately owned, he gave his own money, not that of shareholders.

With still a shortfall of more than £20 million, I managed to pull off what was an original and important deal which has had wide ramifications. Together with Melissa Levitt of the University Development Office, I approached David

Norwood, an entrepreneur and chess grandmaster, who had founded his own company, IndexIT, and backed a number of university spin-outs. The first approach was the crude, 'Give us some money.' In fact, Dave had just sold his company to the London stockbroking firm Beeson Gregory, who had themselves recently had an IPO and were thus quite cash rich and needed to use some of the money raised on their flotation. This was an extraordinary period before the dotcom bubble burst, when financiers were almost fighting to be able to put money into high-tech ventures. Quickly the conversation with Dave moved to thoughts of a deal. His first proposal was that Beeson Gregory would provide some funding in return for the right to be the investors in spin-out companies emanating from the chemistry department. Attractive though that notion was, it was not workable from our point of view. When spin-outs are set up, there has to be the crucial three-way meeting and division of equity involving whoever puts up the money. The university contributes the intellectual property and the academics, without whom it will not work. If one party has the right to be the investor, it would not be possible to bargain on the price. Furthermore, I could not commit my colleagues to accept cash from a guaranteed source. They might prefer to use their own money or that of people who had backed them in earlier ventures. And, of course, chemistry had a few which was why Dave and Beeson Gregory were interested.

The alternative we came up was for the company to provide an upfront sum in return for half of the university's equity in chemistry spin-outs. The parameters were the size of the sum, the percentage of the university's equity (typically 25 per cent), and the length of time of the partnership. After some haggling, a nervous business when playing with a chess grandmaster, we came up with £20 million for half the university equity for a 15-year period.

Having got this deal agreed in principle with Dave, we then had to convince his chairman, Andrew Beeson, of the wisdom of the arrangement. The key meeting was held over a lunch at Gee's in the Banbury Road. The discussion did not start very encouragingly. Andrew seemed unconvinced by the opportunities, but a chance twist in the conversation changed everything. He asked if the tie which I happened to be wearing was a Vincent's Club tie. I replied that it was indeed, and we started to converse about sport, my interests, and then he told me of his interest in real tennis. I responded by asking him whether he had come across one of my pupils who happened to be the world champion under the age of twenty-eight. The whole atmosphere changed. The deal was done, and my pupil, Spike Willcocks, ended up working for Beeson Gregory. The first company from the department which he looked at for Beeson Gregory was Oxford Nanopore Technologies, based on the research of Hagan Bayley. He liked the look of it and jumped ship. Oxford Nanopore, whilst still a private company, has a fair value more than £2 billion.

Surprisingly, the hardest part of the deal was getting the university to accept. The fact that nothing like this had ever been done before counted massively against us. But in the end, a decisive registrar, David Holmes, saw it through. I also had to convince my colleagues that their equity was not affected and that the deal was in their best interests. This was made more complicated by the fact that since Beeson Gregory was a public company, news of the deal was a price-sensitive topic which could not be revealed until the market was informed. I thus had to send out an explanatory email at the same time as the details were announced to the London Stock Exchange and hold a series of meetings with all the chemistry academics. Only one failed to accept the value of the deal. With hindsight, we were extraordinarily lucky with the timing, but in fact all parties have done well out of the arrangement. Chemistry has produced an amazing number of spin-outs with Beeson Gregory providing access to funds as well as business advice, so that the company has had a good return. The academics have set up some fourteen companies, and the university has received over £100 million as a result of spin-outs from the department.

So successful has been the partnership that when Beeson Gregory merged with Evolution, the new group set up a subsidiary with the very dotcom name IP2IPO which made similar deals with ten British universities and was itself floated as an independent public company on the Alternative Investment Market, AIM, before moving to the main London Stock Exchange as the public company IP Group, of which I remained the senior non-executive director for several years, having been for a while the chairman of IP21PO.

The final contribution to the funding for the new laboratory, which we decided to call the Chemistry Research Laboratory, came about as a result of my involvement as a director of IP2IIPO. The company was trying to interest the distinguished American backer of high technology and philanthropist Landon T. Clay in supporting some of their ventures. At the lunch held to encourage this, I happened to end up sitting next to Landon. I remembered that he was the man who had put up five one million dollar prizes for solutions to various mathematical problems. He had been much influenced by the success of Andrew Wiles in solving Fermat's last theorem. He was pretty impressed that I knew that one of his problems was the Birch and Swinnerton-Dyer conjecture. He was not to know that my only acquaintance with this problem was that Bryan Birch happened to be one of my Brasenose colleagues. After the lunch, I went back to my lab and suggested to my secretary, Liz, that we should write a begging letter. We did, and it was successful both in financial terms and in developing a friendship with Landon. Very generously he gave us £250,000 to name one of the laboratories, not after himself, but after Jeremy Knowles, the distinguished Oxford chemist who became head of organic chemistry at Harvard and later dean of arts and sciences there.

In all we raised £64.5 million so that the university had to put not a single penny into the cost of the building. Furthermore we had sufficient funds to do the job properly, without having to cut corners or do anything on the cheap.

On the design I assembled a small, but extremely effective,

committee from amongst my colleagues, one from each chemical tribe, but also with different skills. Jenny Green was invaluable in ensuring the design made sense. Colin Bain virtually designed the basement himself and was adept at reading the plans, while Steve Davies even went so far as to go to Italy to inspect the stone which would be used in the construction.

For the builders, Laing, one of Britain's top, even blue-blooded companies, had been engaged to build the laboratory under a two-stage contract. This means that during the first phase, everything is done with an open book, the design and the preparation of the site. At that stage we, the clients, could sign on with the same builder to proceed to the second stage as was usual, or change to another builder. We were thus more than a little perplexed when, just as we were about to sign up for the second stage, Laing put its construction division up for sale. They had made huge losses on the Millennium Stadium in Cardiff and on the National Physical Laboratory. Despite the predictions of the construction press that the company would be purchased by the major French firm Bouygues, in fact it was sold to the then little-known Ray O'Rourke for the princely sum of £1 with their debt. There was some hesitation amongst the university professionals as to whether we ought to sign for the second stage with the new Laing O'Rourke as they argued that Ray was a concrete man rather than a builder.

To help us decide, I had Ray O'Rourke and some of his senior team to a meeting in my office and was totally won over. I thought he was wonderful and indeed, the way in which he has subsequently almost reinvented the construction industry has proved that we were right.

The construction represented Laing O'Rourke's first major project, and they did a wonderful job for us using some quite novel techniques. Amongst them was using "top-down construction". After a surrounding containing wall was made, again using a novel plastic filter in the trench rather than the traditional Bentonite, the floor plate was put in place. And thereafter, construction went up and down at the same time with consequent saving of time.

The site manager, Mike Morris, did a wonderful job keeping all our neighbours and the Oxford citizenry happy. The site was surrounded by lovely hanging baskets of flowers. The thing which worried me most, however, was that we had to remove 50,000 cubic metres of Oxford clay since much of the building is below ground and, of course, well below the water table. I had visions of massive traffic jams and mud all over the road, for which I would get the blame. In practice, they timed the lorries to avoid rush hours and by radio ensured that there were never queues of trucks. I never received a single complaint.

The only time I did lose my cool was over the date of the final completion. The due date was July 2003, and as that day approached, it was clear that we would not make it. At a meeting in my office, all the professionals swore that they would make 15 September. The certainty of the date was very important to us, not in fact because of the academic year, although that was a consideration, but because once completed, we had to move some very expensive and delicate instruments into the building. Just to move our nuclear magnetic resonance machines was going to require all the Brucker technicians in Europe and had to be planned in advance. I still thought 15 September was ambitious but was assured that this was a certainty. On this basis, we booked the move for 1 October. Come the day they were nowhere near ready, so at serious expense, we had to delay and went for 15 December.

The move had to be made at this time, but in truth, the building was not really ready. The move was massively complex, made worse because one of the few systems which did not perform well were the lifts. We were saved by the fact that one of our better decisions had been to hire the building manager long before construction was complete. Richard Jones was an inspired choice, and it was his skill which ensured that we did get into the building in time for a Royal opening in February 2004.

My original choice to open the building was Margaret Thatcher, an Oxford chemist and the first scientist and first woman premier. Sadly, the vice chancellor vetoed this idea. He was worried that the choice was too controversial, whereas I felt it might undo some of the damage caused by the failure to award her an honorary degree. We fell back on my second choice, Her Majesty the Queen. Happily she was gracious enough to accept the invitation, and we had a wonderful day with her and the Duke of Edinburgh, culminating in a lunch in the large atrium which is a major feature of the laboratory.

Now, after fifteen years of use, it is clear that we have what is possibly the best university chemistry laboratory in the world. In particular, the working conditions for the graduate students could hardly be bettered. They work in a way very reminiscent of a modern pharmaceutical research lab. Each has his or her own workspace in a clean area and can see into the working laboratories through a glass wall.

It was a great pleasure during my last years as chairman taking round distinguished visitors. These included the king of Sweden, the Chinese prime minister, Fidel Castro's son, and many senior academics and industrialists. It was gratifying to hear the man who is probably the most distinguished organic chemist in the world say, 'We have nothing to match this in North America.'

The building as a legacy will obviously last for many years. It was built to the highest standards because we had raised more than sufficient funds. Similarly, the funding company IP Group Plc, that came about as a result of the deal with Dave Norwood, is a permanent benefit now as FTSE 250 company, playing a crucial role in the funding of high-tech companies originating in academic research.

Dave Norwood's next major venture was another leap in imagination, with again important ramifications for the setting up of spin-out companies. He created the company Oxford Sciences Innovation (OSI) which has a novel arrangement with the university whereby the university has "golden shares" in ventures which do not get diluted in subsequent funding rounds. Historically, since universities do not have much spare cash, they were unable to follow this money for further investment rounds and in consequence much diluted.

Dave's important ally in this and in some other investment ventures was Neil Woodford. Currently he is in general not viewed with favour due to the winding up of his investment funds. However, it should not be forgotten that he was hugely important in getting started the investment in UK spin-out companies which has grown to enormous importance.

My own role in getting OSI off the ground was tiny but great fun. Dave appointed me as 'ambassador'. What this meant in the very early days was that very high net worth individuals from the Far East or wherever came to Oxford to discuss with Dave their possible investment in OSI. The habit was for me to give them a tour of the spectacular chemistry research laboratory before they joined Dave for the deal-making. Such are the quality and technical content of the new lab and my enthusiastic tale that when they met with Dave, they were very enthusiastic. More than one offered a £50 million investment. Dave would only accept £15 million, despite their pleading. He was, as ever, very wise. Despite being offered some three billion, he only took in £600 million, still the biggest such fund in the United Kingdom, but not so much that it could not find suitable new companies to support.

In some ways then, with IP Group and OSI, the new laboratory has had an influence on culture, or perhaps even more important, than on just the research in one chemistry department.

12

REFLECTIONS

Much of my scientific life has been quite typical of the academic world as it has developed since the Second World War. I evolved via a doctorate, postdoctoral time, and then a tenured position. I was lucky with timing, coming onstream just as universities worldwide, but especially in the United Kingdom, expanded. I was part of the massive growth in scientific publishing whose commercial publishers joined the learned societies. This was very profitable for many years, including the period when I chaired the publishing side of the Royal Society of Chemistry. Now this is under threat from open-access publishing and hordes of new low-grade journals.

Throughout my scientific life I witnessed the massive growth in international scientific meetings. Many senior academics spend weeks of the year travelling and attending meetings, often in highly desirable locations. Here too, however, there is a disturbing rise in the number of fake and more than dubious events. All established scientists get daily emails touting new journals or conference invitations. These changes are aspects of the underlying commercialisation of science. The impact has not been totally negative, but one does have to be careful and vigilant.

Science prides itself as being truly international. In the recent political storm over the United Kingdom leaving the European Union, a massive majority of scientists were remainers, not just because of the potential loss of EU funding. The growth of China as a scientific powerhouse has been quite phenomenal. In general this has been welcomed, and for pure rather than selfish reasons.

To my mind, however, the biggest change in world science, and one in which I have been much involved, has been the commercialisation of research in the sense of creating new companies or licensing research. When I was participating in this in the early days, it was not looked on with favour. Setting up a company was a naughty thing to do. Now such activity gets credit in research-assessment measures. But it is not without dangers. Any activity where large sums of money are involved runs the danger of dishonesty and conflict. Despite worries that commercialisation might drive scientists into areas overly applied, this does not seem to be a significant problem. As I have mentioned, most of the really successful companies have been derived from pure blue skies research rather than the obviously commercial. It is this rise in the number of new high-technology companies that will provide jobs for today's postdocs and, with care, the development of the economy of the country.

APPENDIX 1

Oxford Research Students

John Horsely	Timothy Walker	Anthony Hall
Reg Hinkley	Alistir Todd	Jim Port
John Raftery	Peter Scott	Les Farnell
Les Clyne	Bob Hammersley	Elizabeth Colbourn
Ian Wilson	Stephen Moore	Tony Marchington
Rachel Klevit	Chet Chung	David Cooper
Valerie Sackwild	Alda Sousa	Robert Elliott
Sandra Robins	Neil Stutchbury	Chris Naylor
Ruth Holmes	Saira Mian	Pippa Bowen-Jenkins
Dave Rickets	Edward Hodgkin	Andrew Smellie
Paul King	Catherine Burt	Garrett Morris
Jonathan Essex	Graham Worth	Richard Gilbert
Paul Boscott	Adrian Elcock	Andrew Good
Adrian Mulholland	Paul Bamborough	David Lewis
Alan Robinson	James Bradley	Thomas Barlow
Stephen Doughty	Ivy Boey	Martin Parretti
Daniel Robinson	Peter Varnai	Owen Walsh
Steward Adcock	Ben Webb	Maya Topf
Ben Allen	Dave Huggens	Chris Baker

Postdoctoral Researchers

S. P. So	Jill Gready	Harit Trivedi
Roger Humphries	George Jaraskiewiz	Alastair Cuthbertson
Chris Reynolds	Ian Haworth	Guy Grant
Barry Hardy	Jeff Rothman	Paul Lyne
Romano Kroemer	Ana Castro	Peter Winn
Thomas Barlow	Aaron Dinar	Xabier Lopez
Meir Glick	Walter Scott	Aixia Yan
Meilan Huang	Pedro Ballester	Loris Morretti

Sabbatical Visitors

Juan Adelantado, Massoud Mahmoudian, Vijay Gomba, Akira Nakayama, Myrna Gil, Vernon Cheney, Federico Crago, Cristina Menziani, Gyorygi Ferenczy, Amatz Meyer, Maria Ramos, Carl Schwalbe, Aron Seri-Levy, Sayaka Shinomoto, Justin Caravella, Anderson Coser Gaudio, Perter Varnai, Tony Hopfinger, Dave Winkler, Guy Grant, Milan Remko.

To all of the people mentioned above, I owe an enormous debt of gratitude. They are the folk who did the hard work.

APPENDIX 2

PUBLICATIONS

Books

1. *Ab Initio Molecular Orbital Calculations for Chemists.* Clarendon Press, Oxford 1970 (with J. A. Horsley). 2nd Edition 1983 (with D. L. Cooper).
2. *Bibliography of Ab Initio Molecular Wavefunctions.* Clarendon Press, Oxford 1971 (with T. E. H. Walker and R. K. Hinkley).
3. *Supplement 1970–73.* Clarendon Press, Oxford 1974 (with T. E. H. Walker, L. Farnell, and P. R. Scott).
4. *Supplement 1974–77.* Clarendon Press, Oxford 1978 (with P. R. Scott, E. A. Colbourn, and A. F. Marchington).
5. *Supplement 1978–80.* Clarendon Press, Oxford 1981 (with P. R. Scott, V. Sackwild, and S. A. Robins).
6. *Entropy and Energy Levels.* Clarendon Press, Oxford 1974 (with R. P. H. Gasser); Spanish edition 1978.
7. *Structure and Spectra of Atoms.* John Wiley, Chichester 1976 (with P. R. Scott); Japanese edition 1996.

8. *Quantum Pharmacology.* Butterworth, London 1977. 2nd edition 1983; Japanese edition 1986.
9. *Spin-Orbit Coupling in Molecules.* Oxford University Press 1981
   (with H. P. Trivedi and D. L. Cooper).
10. *Structure and Spectra of Molecules.* John Wiley, Chichester 1985 (with P. R. Scott).
11. *Oxford Illustrated Encyclopaedia. Vol. I: The Physical World* (editor). Oxford University Press 1985.
12. *The Problems of Chemistry.* Oxford University Press 1986.
    French edition 1989; German edition 1990;
    Spanish edition 1991; Japanese edition 1991.
13. *Computer-Aided Molecular Design* (editor). IBC London; VCH, New York 1989.
14. *Energy Levels of Atoms and Molecules.*
    Oxford University Press 1994 (with P. R. Scott).
    Persian edition 2001.
15. *Computational Chemistry* Oxford University Press 1995 (with G. H. Grant).
16. *An Introduction to Statistical Thermodynamics.*
    World Scientific Publishing Co. 1995 (with R. P. H. Gasser).
17. *Spin-outs. Creating Businesses from University Intellectual Property.*
    Harriman House 2009.
18. *50 Years at Oxford.*
    AuthorHouse 2011.

19. *University Intellectual Property: A Source of Finance and Impact.*
Harriman-House 2012.
20. *Entrepreneurship: A Case Study from Two View Points.*
WetZebra 2016.

The Evans family

Gregynog Hall

BA graduation

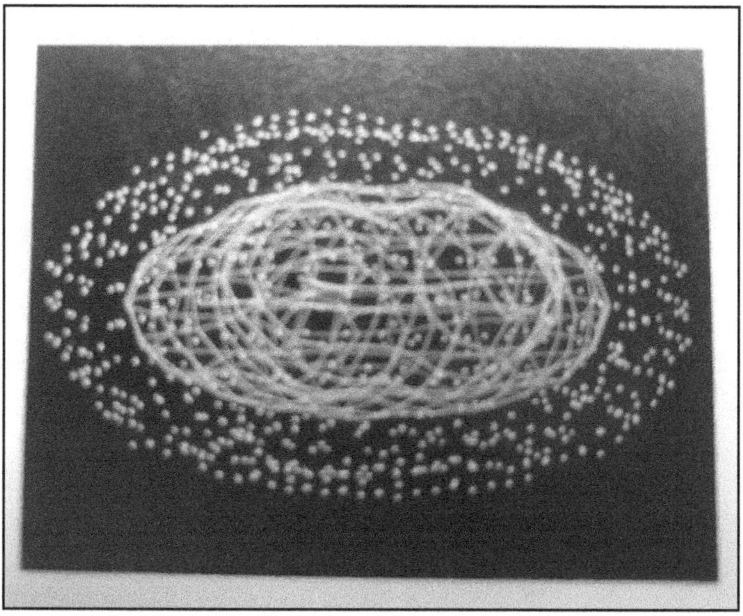

Early computer graphics

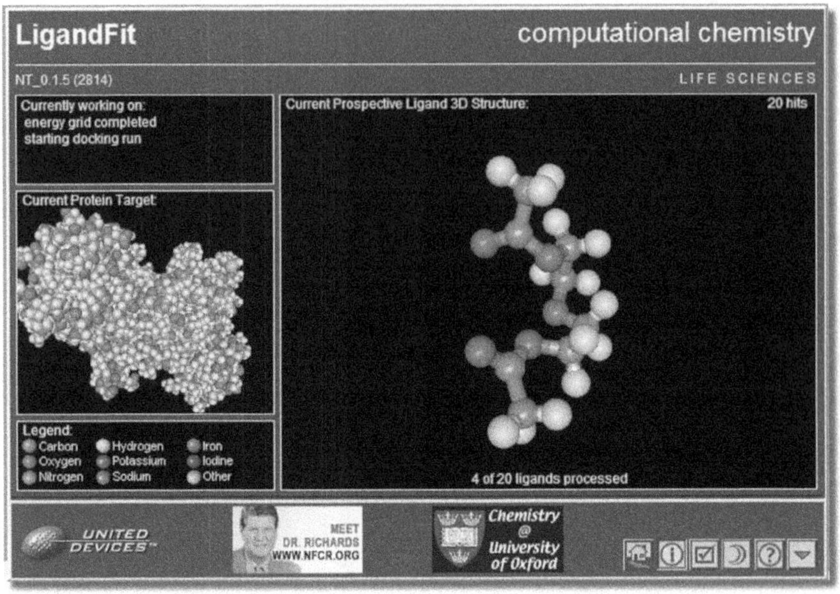

The screensaver

Wen Jiabao visiting the Laboratory

The new Laboratory

HM Thé Queen opening the new Laboratory

Lightning Source UK Ltd.
Milton Keynes UK
UKHW040620240121
377563UK00001B/21

9 781665 584449